Selected Titles in This Series

607 **Adele Zucchi,** Operators of class C_0 with spectra in multiply connected regions, 1997

606 **Moshé Flato, Jacques C. H. Simon, and Erik Taflin,** Asymptotic completeness, global existence and the infrared problem for the Maxwell-Dirac equations, 1997

605 **Liangqing Li,** Classification of simple C^*-algebras: Inductive limits of matrix algebras over trees, 1997

604 **Hajnal Andréka, Steven Givant, and István Németi,** Decision problems for equational theories of relation algebras, 1997

603 **Bruce N. Allison, Saeid Azam, Stephen Berman, Yun Gao, and Arturo Pianzola,** Extended affine Lie algebras and their root systems, 1997

602 **Igor Fulman,** Crossed products of von Neumann algebras by equivalence relations and their subalgebras, 1997

601 **Jack E. Graver and Mark E. Watkins,** Locally finite, planar, edge-transitive graphs, 1997

600 **Ambar Sengupta,** Gauge theory on compact surfaces, 1997

599 **Tai-Ping Liu and Yanni Zeng,** Large time behavior of solutions for general quasilinear hyperbolic-parabolic systems of conservation laws, 1997

598 **Valentina Barucci, David E. Dobbs, and Marco Fontana,** Maximality properties in numerical semigroups and applications to one-dimensional analytically irreducible local domains, 1997

597 **Ragnar-Olaf Buchweitz and John J. Millson,** CR-geometry and deformations of isolated singularities, 1997

596 **Paul S. Bourdon and Joel H. Shapiro,** Cyclic phenomena for composition operators, 1997

595 **Eldar Straume,** Compact connected Lie transformation groups on spheres with low cohomogeneity, II, 1997

594 **Solomon Friedberg and Hervé Jacquet,** The fundamental lemma for the Shalika subgroup of $GL(4)$, 1996

593 **Ajit Iqbal Singh,** Completely positive hypergroup actions, 1996

592 **P. Kirk and E. Klassen,** Analytic deformations of the spectrum of a family of Dirac operators on an odd-dimensional manifold with boundary, 1996

591 **Edward Cline, Brian Parshall, and Leonard Scott,** Stratifying endomorphism algebras, 1996

590 **Chris Jantzen,** Degenerate principal series for symplectic and odd-orthogonal groups, 1996

589 **James Damon,** Higher multiplicities and almost free divisors and complete intersections, 1996

588 **Dihua Jiang,** Degree 16 Standard L-function of $GSp(2) \times GSp(2)$, 1996

587 **Stéphane Jaffard and Yves Meyer,** Wavelet methods for pointwise regularity and local oscillations of functions, 1996

586 **Siegfried Echterhoff,** Crossed products with continuous trace, 1996

585 **Gilles Pisier,** The operator Hilbert space OH, complex interpolation and tensor norms, 1996

584 **Wayne W. Barrett, Charles R. Johnson, and Raphael Loewy,** The real positive definite completion problem: Cycle completability, 1996

583 **Jin Nakagawa,** Orders of a quartic field, 1996

582 **Darryl McCollough and Andy Miller,** Symmetric automorphisms of free products, 1996

581 **Martin U. Schmidt,** Integrable systems and Riemann surfaces of infinite genus, 1996

580 **Martin W. Liebeck and Gary M. Seitz,** Reductive subgroups of exceptional algebraic groups, 1996

579 **Samuel Kaplan,** Lebesgue theory in the bidual of $C(X)$, 1996

578 **Ale Jan Homburg,** Global as[illegible] 1996

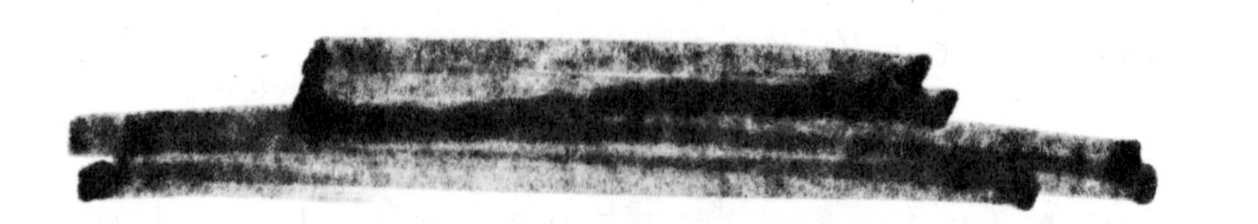

Operators of Class C_0 with Spectra in Multiply Connected Regions

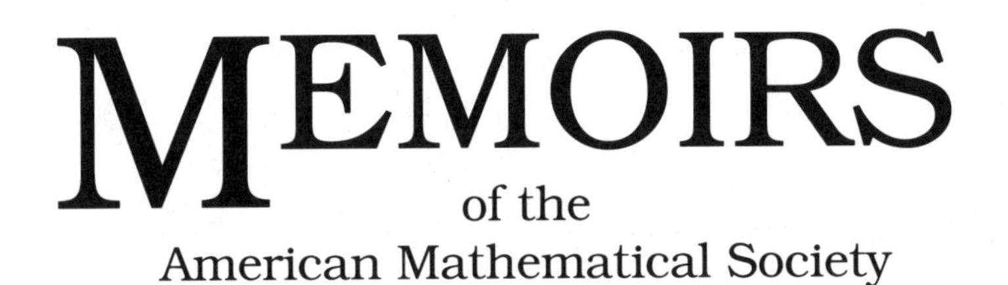

Number 607

Operators of Class C_0 with Spectra in Multiply Connected Regions

Adele Zucchi

May 1997 • Volume 127 • Number 607 (third of 4 numbers) • ISSN 0065-9266

American Mathematical Society
Providence, Rhode Island

1991 *Mathematics Subject Classification.*
Primary 47A45; Secondary 47A60, 30D55.

Library of Congress Cataloging-in-Publication Data

Zucchi, Adele, 1966–
Operators of class C_0 with spectra in multiply connected regions / Adele Zucchi.
p. cm. — (Memoirs of the American Mathematical Society, ISSN 0065-9266 ; no. 607)
"May 1997, volume 127, number 607 (third of 4 numbers)."
Includes bibliographical references.
ISBN 0-8218-0626-2 (alk. paper)
1. Nonselfadjoint operators. 2. Functional analysis. 3. Hardy classes. I. Title. II. Series.
QA3.A57 no. 607
[QA329.2]
510 s—dc21
[515′.7246]

97-3959
CIP

Memoirs of the American Mathematical Society

This journal is devoted entirely to research in pure and applied mathematics.

Subscription information. The 1997 subscription begins with number 595 and consists of six mailings, each containing one or more numbers. Subscription prices for 1997 are $414 list, $331 institutional member. A late charge of 10% of the subscription price will be imposed on orders received from nonmembers after January 1 of the subscription year. Subscribers outside the United States and India must pay a postage surcharge of $30; subscribers in India must pay a postage surcharge of $43. Expedited delivery to destinations in North America $35; elsewhere $110. Each number may be ordered separately; *please specify number* when ordering an individual number. For prices and titles of recently released numbers, see the New Publications sections of the *Notices of the American Mathematical Society.*

Back number information. For back issues see the *AMS Catalog of Publications.*

Subscriptions and orders should be addressed to the American Mathematical Society, P. O. Box 5904, Boston, MA 02206-5904. *All orders must be accompanied by payment.* Other correspondence should be addressed to Box 6248, Providence, RI 02940-6248.

Memoirs of the American Mathematical Society is published bimonthly (each volume consisting usually of more than one number) by the American Mathematical Society at 201 Charles Street, Providence, RI 02904-2294. Periodicals postage paid at Providence, RI. Postmaster: Send address changes to Memoirs, American Mathematical Society, P. O. Box 6248, Providence, RI 02940-6248.

This publication is indexed in *Science Citation Index*®, *SciSearch*®, *Research Alert*®, *CompuMath Citation Index*®, *Current Contents*®/*Physical, Chemical & Earth Sciences.*
Printed in the United States of America.

♾ The paper used in this book is acid-free and falls within the guidelines established to ensure permanence and durability.

10 9 8 7 6 5 4 3 2 1   02 01 00 99 98 97

Contents

1. Introduction **1**

2. Preliminaries and Notation **4**
2.1. Contractions of class C_0 4
2.2. $H^p(\Omega)$ Spaces 5
2.3. Arithmetic of inner functions in $H^\infty(\Omega)$ 9

3. The Class C_0 **15**
3.1. Construction of the functional calculus 15
3.2. The class C_{00} 19
3.3. Minimal functions and maximal vectors 21
3.4. General properties of the class C_0 24

4. Classification Theory **29**
4.1. Jordan blocks 29
4.2. Multiplicity-free operators 34
4.3. The classification theorem 38

Bibliography **51**

Abstract. Let Ω be a bounded finitely connected region in the complex plane, whose boundary Γ consists of disjoint, analytic, simple closed curves. We consider linear bounded operators on a Hilbert space H having $\overline{\Omega}$ as spectral set, and no normal summand with spectrum in Γ. For each operator satisfying these properties, we define a weak*-continuous functional calculus representation $\Phi : H^\infty(\Omega) \to \mathcal{L}(H)$, where $H^\infty(\Omega)$ is the Banach algebra of bounded analytic functions on Ω. An operator is said to be of class C_0 if the associated functional calculus has a non-trivial kernel. In this paper we study operators of class C_0, for which we provide a complete classification into quasisimilarity classes, analogous to the case of the unit disk.

Key words and phrases. Hardy spaces; Hilbert spaces; Functional Calculus; Classification; Quasisimilarity; Jordan blocks.

Author address:

Dipartimento di Matematica - Università di Milano
via Saldini 50, 20133 Milano, ITALY
E-mail address: zucchi@vmimat.mat.unimi.it

Received by the editor December 12, 1994.

1. Introduction

One branch of operator theory which has blossomed during the last four decades is the study of Hilbert spaces operators related to the unit disk $D = \{z \in \mathbf{C} : |z| < 1\}$. Although its origin can be traced to the von Neumann-Wold characterization of isometry [28, 32], the subject began in earnest with Beurling's determination of the invariant subspaces of the shift operator [10] and von Neumann's work on spectral sets [29]. In [24] B.Sz.-Nagy gave an alternative proof of von Neumann's theorem that the unit disk is a spectral set for contractions via unitary dilations and then developed the latter notion in collaboration with C. Foias into a model theory for contraction operators [27]. Operators of class C_0 with spectrum in the closed unit disk were introduced by B.Sz.-Nagy and C. Foias in their work on canonical models for contractions. A completely nonunitary contraction belongs to this class if the associated functional calculus on H^∞ has a non-trivial kernel. Since when the class C_0 was first defined [24], it was studied by many others (see [7]) and, quite possibly, is the best understood class of non-normal operators.

In this paper Ω will be a bounded finitely connected region in the complex plane, whose boundary Γ consists of disjoint, analytic, simple closed curves. Let $R(\Omega)$ be the space of rational functions with poles off $\overline{\Omega}$, and $\mathrm{Rat}(\overline{\Omega})$ be the closure in $C(\overline{\Omega})$ of $R(\Omega)$. Let H be a complex Hilbert space and let $\mathcal{L}(H)$ be the algebra of bounded linear operators on H. M.B. Abrahamse and R.G. Douglas [4] initiated in 1974 the study of contractive unital $\mathcal{L}(H)$-representation of $\mathrm{Rat}(\overline{\Omega})$. In this paper and two years later in a paper about subnormal operators related to multiply connected regions [5], they quoted a preprint in preparation about C_0 operators over multiply connected regions, that was never published. In 1978 in a paper about operators of class C_{00} over multiply connected regions [6] J.A. Ball refers to operators of class C_0 for finitely connected regions saying without proof that an operator of class C_0 is of class C_{00} too. Abrahamse and Douglas were planning to develop the theory of the class C_0 starting from their study of bundle shifts [5]. Our approach here avoids almost completely the use of bundle shifts. Bundle shifts would be necessary in a theory where unitary equivalence is the basic classification criterium, while we only consider similarity and quasisimilarity.

We consider linear bounded operators on a Hilbert space having $\overline{\Omega}$ as spectral set, and no normal summand with spectrum in Γ. For each operator satisfying these properties, we define a weak*-continuous functional calculus $\Phi : H^\infty(\Omega) \to \mathcal{L}(H)$, where $H^\infty(\Omega)$ is the Banach algebra of bounded analytic functions on Ω (under somewhat more restrictive hypotheses this functional calculus was considered earlier by B. Chevreau, C.M. Pearcy and A.L. Shields [11]). An operator is said to be of class C_0 if the associated functional calculus has a non-trivial kernel. The central object of this paper are operators of class C_0, for which we provide a complete classification into quasisimilarity classes analogous to Jordan's classical results in finite-dimensional linear algebra. Our work can be viewed as an extension to more general regions of results first proved for the case of the disk.

The influence of Sz.-Nagy Dilation Theorem on the development of operator theory has been extraordinary. Indeed, it removed much of the mistery that had surrounded von Neumann's Inequality by establishing a simple geometric explanation. Moreover it prompted a number of mathematicians to ask the following question:
Let $F \subseteq \mathbf{C}$ be a compact set and let $T \in \mathcal{L}(H)$. Are the following two conditions equivalent?

(i) F is a spectral set for T.

(ii) There exists a Hilbert space $K \supseteq H$ and a normal operator $N \in \mathcal{L}(K)$ such that $\sigma(N) \subseteq \partial F$ and if P_H denotes the orthogonal projection of K onto H, then $f(T) = P_H f(N)_{|H}$, for all $f \in \mathrm{Rat}(F)$.

In [12] R.C. Douglas and V. Paulsen show that if F is sufficiently "nice" (roughly, $\mathrm{Rat}(F)$ is hypo-Dirichlet) and $S \in \mathcal{L}(H)$ has F as spectral set, then S is similar to an operator $T \in \mathcal{L}(H)$ satisfying (ii). In [2] J. Agler proves that the answer to this question is yes if F is an annulus, but for general compact sets F the question remains unsolved to this day. For this reason we can not use the powerful tool of the existence of a normal dilation.

An important tool for our work, and in general for the study of Hilbert space operators related to multiply connected regions, is the characterization of fully invariant subspaces of $H^p(\Omega)$. This characterization closely resembles Beurling's characterization of the shift invariant subspaces for the Hardy spaces on the disk. Sarason [22] provided the description in the case where Ω is an annulus, and Hasumi [16] and Voichick [30, 31] for more general regions.

What makes C_0 operators so well-understood is that their properties are closely related with the arithmetic of $H^\infty(\Omega)$. Our work is therefore partially devoted to the stủdy of functions in $H^\infty(\Omega)$. There are two approaches to function theory on multiply connected regions. The first is to work directly on the region and examine the analytic functions in their own customary setting. The second approach is to "lift" the function theory of the region to the unit disk by means of a covering projection map. This technique has clear advantages, but new difficulties arise with the requirement that the functions must be invariant under the group of linear fractional transformations that fix the covering map. We adopt the first approach, which we find more suitable for our purposes.

We use single-valued holomorphic functions. In order to accomplish this, we shall often need to insert into our formulae harmonic functions continuous up to the boundary Γ and constant on each boundary component. This will mean that inner functions, etc., are only required to have moduli which are constant almost everywhere on each boundary component of Ω, rather than having those which are one almost everywhere on Γ. Sarason [23], Hasumi [16], Voichick [30, 31] and others take a different approach, and allow their functions to be multivalued, but restrict inner functions, etc., to those whose boundary values are 1 in modulus almost everywhere.

This paper consists of four chapters, the first of which is the introduction. In the second we recall some of the results about operators of class C_0 over the unit disk. We only give a short summary of the results we need, since most of them will be generalized in chapters 3 and 4. Then we present Hardy spaces on general planar regions. Lastly we discuss certain properties of $H^\infty(\Omega)$ that will be essential in what follows. First we explain the generalization of the concepts of divisibility, greatest common inner divisor and least common inner multiple to functions over multiply connected regions; then we deal with some analytical properties of inner functions in order to define the concept of support of inner functions, in analogy with the same concept for inner functions on the disk.

The third chapter contains the theory of operators of class C_0. We introduce the class and study its elementary properties related to adjoints, invariant subspaces and functional calculus. We show that operators of class C_0 have a minimal inner function, which is analogous to the minimal polynomial of a finite matrix, and that every inner function occurs as the minimal function of some operator of class C_0. Then we define operators of class C_{00} and we show that an operator of class C_0 is of class C_{00}. Operators of class C_0 admit a "local" characterization in the sense that an operator is of class C_0 if and only if its restriction to each rationally cyclic invariant subspace (i.e., of the form $\bigvee\{r(T)x : r \in R(\Omega)\}$ for some $x \in H$), is of class C_0. Finally the last part contains more properties related to this class of operators, namely a relation between minimal function and spectrum and relations between minimal functions and rationally invariant subspaces.

A Jordan operator is a direct sum of Jordan blocks with certain additional properties, a Jordan block is the compression of the operator of multiplication by z on $H^2(\Omega)$ to the orthogonal complement of a fully invariant subspace. In the last chapter we achieve a complete characterization of operators of class C_0 by showing that each quasisimilarity class contains a unique Jordan operator. We start by studying operators of class C_0 having a rationally cyclic vector, also called multiplicity-free operators. It turns out that each such operator is quasisimilar to a unique Jordan block, and this is a first step in the classification. Then we extend the classification results to operators with higher multiplicity. Many of the ideas and contents of section 4.3 are similar in structure to those in [7] for the case of the disk. Those proofs which are very similar to the proofs available in the case of the disk are only briefly sketched.

This paper is based on the author's Indiana University Doctoral Dissertation, submitted in 1994. The author would like to express her gratitude to her thesis adviser, Professor Hari Bercovici.

2. Preliminaries and Notation

2.1. Contractions of class C_0. If $\{\mathcal{M}_i\}_{i\in I}$ is a family of subsets of the Hilbert space H, we denote by $\bigvee_{i\in I}\mathcal{M}_i$ the closed linear span generated by $\bigcup_{i\in I}\mathcal{M}_i$. Moreover, $\mathcal{M}^-$ denotes the closure of $\mathcal{M}$, for any subset $\mathcal{M}$ of H.

DEFINITION 2.1.1. A contraction $T \in \mathcal{L}(H)$ is said to be *completely nonunitary* if there is no invariant subspace $\mathcal{M}$ for T such that $T_{|\mathcal{M}}$ is a unitary operator.

If K is a Hilbert space, $H \subseteq K$ is a subspace, $S \in \mathcal{L}(K)$ and $T \in \mathcal{L}(H)$, then we say that S is a *dilation* of T (and T is a *power-compression* of S) provided that $T^n = P_H S^n_{|H}$ for $n \in \{0, 1, \dots\}$.

If in addition S is an isometry (unitary operator) then S will be called an *isometric (unitary) dilation* of T. S is a *minimal* isometric (unitary) dilation of T if and only if $\bigvee_{n=0}^{\infty} S^n H = K$ ($\bigvee_{n=-\infty}^{\infty} S^n H = K$).

THEOREM 2.1.2. **Sz.-Nagy Dilation Theorem**([24]) *(i) Every contraction $T \in \mathcal{L}(H)$ has a minimal isometric dilation. This dilation is unique in the following sense: if $S \in \mathcal{L}(K)$ and $S' \in \mathcal{L}(K')$ are two minimal isometric dilations for T, then there exists an isometry U of K onto K' such that $Ux = x$, $x \in H$, and $S'U = US$.*
(ii) Every contraction $T \in \mathcal{L}(H)$ has a minimal unitary dilation. This dilation is unique in the sense specified in (i).

THEOREM 2.1.3. ([25]) *The spectral measure of the minimal unitary dilation of a completely nonunitary contraction is mutually absolutely continuous with respect to arc-length measure on the unit circle* $\mathbf{T}$.

Let $T \in \mathcal{L}(H)$ be a completely nonunitary contraction with minimal unitary dilation $U \in \mathcal{L}(K)$, and let P_H denote the orthogonal projection onto H. For every polynomial $p(z) = \sum_{j=o}^{n} a_j z^j$ we then have $p(T) = P_H p(U)_{|H}$, and this relation suggests that the functional calculus $p \to p(T)$ might be extended to more general functions. More precisely, since the spectral measure of U is absolutely continuous with respect to arc-length on $\mathbf{T}$, the expression $f(U)$ makes sense for every f in $L^\infty(\mathbf{T})$. Therefore it is possible to define $f(T)$ by $f(T) = P_H f(U)_{|H}$ for all $f \in L^\infty(\mathbf{T})$. Even if the mapping $f \to f(T)$ is obviously linear, it is not in general multiplicative. It turns out that there is a unique maximal algebra such that, for all operators T, the map is multiplicative. This algebra is H^∞, i.e., the algebra of bounded analytic functions on D.

A *representation* of H^∞ into $\mathcal{L}(H)$ is an algebra homomorphism of H^∞ into $\mathcal{L}(H)$. In the following proposition we mention some important properties of this functional calculus [27].

PROPOSITION 2.1.4. *Let $T \in \mathcal{L}(H)$ be a completely nonunitary contraction. Then there exists a unique representation Φ of H^∞ into $\mathcal{L}(H)$ such that:*
(i) $\Phi(\mathbf{1}) = \mathbf{1}_H$, where $\mathbf{1}_H \in \mathcal{L}(H)$ is the identity operator and $\mathbf{1}(z) = 1$ for all $z \in D$;
(ii) $\Phi(g) = T$, where $g(z) = z$ for all $z \in D$;
(iii) Φ is continuous when H^∞ and $\mathcal{L}(H)$ are given the weak-topology.*
We also have that Φ is contractive, i.e. $\|\Phi(u)\| \leq \|u\|$ for all $u \in H^\infty(\Omega)$.

We simply denote by $u(T)$ the operator $\Phi(u)$.

PROPOSITION 2.1.5. *If for every $u \in H^\infty$ we denote by $\tilde{u} \in H^\infty$ the function defined by $\tilde{u}(z) = \overline{u(\overline{z})}$, we then have $\tilde{u}(T)^* = u(T^*)$.*

DEFINITION 2.1.6. A completely nonunitary contraction $T \in \mathcal{L}(H)$ is said to be of *class C_0* if there exists $u \in H^\infty$, $u \neq 0$, such that $u(T) = 0$.

Let T be an operator of class C_0. Then the set $J = \{u \in H^\infty : u(T) = 0\}$ is a weak*-closed ideal, and hence it is of the form $J = vH^\infty$ for some inner function v ([10]).

DEFINITION 2.1.7. The inner function v such that $vH^\infty = \{u \in H^\infty : u(T) = 0\}$ is called the *minimal function* of T and is denoted by m_T.

Let us note that the function m_T is determined only up to a constant scalar multiple of absolute value one.

DEFINITION 2.1.8. A contraction $T \in \mathcal{L}(H)$ is said to be of class C_{*0} if for all $x \in H$ $\lim_{n\to\infty} \|T^{*^n} x\| = 0$; T is of class C_{0*} if T^* is of class C_{*0}. Finally, T is of class C_{00} if it is both of class C_{*0} and C_{0*}.

PROPOSITION 2.1.9. ([26]) *If T is of class C_0, then it is of class C_{00}.*

2.2. $H^p(\Omega)$ Spaces. The theory of Hardy spaces over multiply connected regions has been first studied by Rudin [21]; cf. also [13]. Much of what is known for the unit disk D extends to a region whose boundary consists of finitely many disjoint, analytic, simple closed curves, since many questions can be reduced to the simply connected case by means of a decomposition theorem.

Let $1 \leq p < \infty$. A holomorphic function f on Ω is in $H^p(\Omega)$ if the subharmonic function $|f|^p$ has a harmonic majorant on Ω. For a fixed $z_0 \in \Omega$, there is a norm on $H^p(\Omega)$ defined by

$$\|f\| = \inf\{u(z_0)^{1/p} : u \quad \text{is a harmonic majorant of} \quad |f|^p\}.$$

Let $g(\cdot, z_0)$ be the Green's function for Ω with pole at z_0, and ω be the harmonic measure on Γ for the point z_0. In our case, since the region Ω has a nice boundary, the harmonic measure is easy to understand as the following result ([13] Theorem 1.6.4) shows.

THEOREM 2.2.1. *Let z_0 be any point in Ω. Then we have that $d\omega = \frac{-1}{2\pi}\frac{\partial}{\partial n} g(\cdot, z_0) ds$, where $\frac{\partial}{\partial n}$ is the derivative in the direction of the outward normal at Γ, and ds is the arc-length measure on Γ.*

The function $\frac{-1}{2\pi}\frac{\partial}{\partial n}g(\zeta, z_0)$, $\zeta \in \Gamma$ and $z_0 \in \Omega$, is the analog for Ω of the Poisson kernel for D (Green's kernel). It is possible to show ([13] Proposition 1.6.6) that $\frac{\partial}{\partial n}g(\zeta, z_0) < 0$ for $\zeta \in \Gamma$. As an immediate consequence of this fact, we have that $d\omega$ is mutually absolutely continuous with respect to ds for all $z_0 \in \Omega$.

Each function $f \in H^p(\Omega)$ has nontangential boundary values f^* almost everywhere $d\omega$, and the function f^* defined by these limits is in $L^p(\Gamma, \omega)$. The mapping $f \to f^*$ is an isometry from $H^p(\Omega)$ onto a closed subspace of $L^p(\Gamma, \omega)$. Further if the boundary values f^* vanish on a set of positive measure, so does the function f.

A function f defined on Ω is in $H^\infty(\Omega)$ if it is holomorphic and bounded. $H^\infty(\Omega)$ is a closed subspace of $L^\infty(\Gamma, \omega)$ and it is a Banach algebra if endowed with the supremum norm. Finally, the mapping $f \to f^*$ is an isometry of $H^\infty(\Omega)$ onto a weak*-closed subalgebra of $L^\infty(\Gamma, \omega)$.

Let $1 \leq p \leq \infty$. We will follow the common practice of using the same letter f to denote both the function on Ω and its boundary values. In this way $H^p(\Omega)$ can be viewed as a closed subspace of $L^p(\Gamma, \omega)$. In particular, $H^\infty(\Omega)$ can be viewed as a weak*-closed subalgebra of $L^\infty(\Gamma, \omega)$, and $H^2(\Omega)$ is a Hilbert space. Further, if $\Omega = D$, we denote $H^p(D)$ simply by H^p.

The classes $H^p(\Omega)$ are invariant under conformal transformations of Ω. Let ϕ be a one-to-one holomorphic mapping of a region Ω onto a region Ω'. Then the map $V : H^p(\Omega') \to H^p(\Omega)$ defined by $Vf = f \circ \phi$, is a bounded invertible operator. If the $H^p(\Omega')$ norm is determined at the point $\phi(z_0)$, then V is also an isometry.

We recall that a sequence $\{f_n\}_{n=1}^\infty$ of elements in $H^\infty(\Omega)$ is weak*-convergent if and only if it is boundedly pointwise convergent, i.e., it converges pointwise and $\sup_{n\in\mathbf{N}} \|f_n\|_\infty$ is finite.

PROPOSITION 2.2.2. ([14]) *$R(\Omega)$ is sequentially boundedly weak*-dense in $H^\infty(\Omega)$, i.e., for any $f \in H^\infty(\Omega)$ there exists a sequence $\{r_n\}_{n=1}^\infty$ in $R(\Omega)$ weak*-convergent to f and such that* $\max\{|r_n(z)| : z \in \overline{\Omega}\} \leq \|f\|_\infty$ *for all n.*

We continue this section by reformulating (following [20]) the concepts of inner and outer functions and of Blaschke product and singular inner function in a manner suitable for function theory on multiply connected regions, so that the classical factorization theorems remain true.

Let us first recall that if h is a real-valued harmonic function defined in Ω and $d^*h = (\partial h/\partial x)dy - (\partial h/\partial y)dx$, then, for $j = 1, \dots, m$, the number $\int_{\Gamma_j} d^*h$ is called the *period* about Γ_j of the harmonic conjugate ${}^*h$ (or of the conjugate differential d^*h) of h. There are many times when it is highly advantageous to "correct" these periods, that is, to add a harmonic function h_0 to h so that the periods of ${}^*(h + h_0)$ are all zero, and thus ${}^*(h + h_0)$ is single-valued. There are several ways to do this (cf. [13, 19]). In particular for any harmonic function h on Ω there exists a harmonic function h_0 on Ω which is continuous on $\overline{\Omega}$, constant on each component of Γ and such that $h + h_0$ has a single-valued harmonic conjugate ([18] p.304).

DEFINITION 2.2.3. A nonzero function θ in $H^\infty(\Omega)$ is said to be *inner* if $|\theta|$ is constant almost everywhere on each component of Γ.

A function F in $H^p(\Omega)$ is said to be *outer* if for all $z \in \Omega$ we have

$$\log|F(z)| = -\frac{1}{2\pi}\int_\Gamma \log|F(\zeta)|\frac{\partial g}{\partial n}(\zeta, z)ds.$$

An inner function is said to be *trivial* if it is invertible. The invertible inner functions form a finitely generated group under multiplication. An outer function has no zeros. The product of two inner (or outer) functions is inner (or outer), and so is their quotient if it is bounded. An inner function is outer if and only if it is trivial.

It is known [13, 20] that the factorization of a function in H^p into a product of an inner and an outer function carries over to $H^p(\Omega)$. The following result follows from Theorem 4.7.3 in [13].

THEOREM 2.2.4. *Each function $f \in H^p(\Omega)$ has a factorization $f = \theta F$, where θ is inner and F is an outer function in $H^p(\Omega)$. If f is not identically zero, then θ and F are uniquely determined up to a trivial inner factor.*

DEFINITION 2.2.5. Given two inner functions θ and θ', we say that θ is *equivalent* to θ' ($\theta \equiv \theta'$) if there exists a trivial inner function ψ such that $\theta = \psi\theta'$.

Then, if $\mathbf{1}$ denotes the constant function in $H^2(\Omega)$ with values equal to 1, θ is trivial if and only if $\theta \equiv \mathbf{1}$.

DEFINITION 2.2.6. An inner function S with no zeros is called a *singular* function.

THEOREM 2.2.7. *(i) Given any positive measure ν on Γ singular relative to arc-length, there exists a unique (up to equivalence) singular function S_ν such that for all $z \in \Omega$*

$$\log|S_\nu(z)| = \int_\Gamma \frac{\partial g}{\partial n}(\zeta, z)d\nu(\zeta) + h(z),$$

where h is a harmonic function on Ω, continuous on $\overline{\Omega}$ with constant values on every component of Γ.
(ii) If S is a singular function, then there exist a unique positive measure ν on Γ singular relative to arc-length, such that $S \equiv S_\nu$.

We call ν the *representing measure* of S_ν.

It is known ([13] Proposition 4.7.1) that if a function in $H^\infty(\Omega)$ is not identically zero, and if $\{a_n\}_{n=1}^\infty$ is the sequence of its zeros repeated according to multiplicity, then for each $z \in \Omega$ we have $\sum_{n=1}^\infty g(z, a_n) < \infty$, and the convergence is uniform on compact subsets of $\Omega - \{a_n\}_{n=1}^\infty$. Moreover ([18] Theorem 3.1) the uniform convergence on compact subsets of $\Omega - \{a_n\}_{n=1}^\infty$ of $\sum_{n=1}^\infty g(z, a_n)$ is equivalent to the convergence of $\sum_{n=1}^\infty \operatorname{dist}(a_n, \Gamma)$.

DEFINITION 2.2.8. A function $\mu : \Omega \to \mathbf{N} = \{0, 1, 2, \dots\}$ is said to be a *Blaschke function* if the series $\sum_{a\in\Omega} \mu(a) \operatorname{dist}(a, \Gamma)$ is convergent.

THEOREM 2.2.9. *If μ is a Blaschke function, then there exists a unique (up to equivalence) inner function B_μ having at each $a \in \Omega$ a zero with multiplicity $\mu(a)$, and such that for all $z \in \Omega$*

$$\log |B_\mu(z)| = -\sum_{a\in\Omega} \mu(a)g(z,a) + h(z),$$

where h is a harmonic function on Ω, continuous on $\overline{\Omega}$ with constant values on every component of Γ.

We call μ the *representing Blaschke function* of B_μ.

DEFINITION 2.2.10. The functions B_μ defined in Theorem 2.2.9 are called *(generalized) Blaschke products.*

Clearly the product of two Blaschke products is a Blaschke product, and so is their quotient if it is bounded.

THEOREM 2.2.11. ([13]) *Any inner function θ may be factored into $\theta = BS$, where B is a Blaschke product having the same zeros of θ, and S is a singular function. The factors are unique up to equivalence.*

LEMMA 2.2.12. *(i) If S_ν and $S_{\nu'}$ are singular functions, then $S_\nu S_{\nu'} \equiv S_{\nu+\nu'}$. (ii) If B_μ and $B_{\mu'}$ are Blaschke products, then $B_\mu B_{\mu'} \equiv B_{\mu+\mu'}$.*

DEFINITION 2.2.13. A closed linear subspace $\mathcal{M}$ of $H^p(\Omega)$ (weak*-closed if $p = \infty$) is said to be *fully invariant* if $rf \in \mathcal{M}$ for all $f \in \mathcal{M}$ and for all $r \in R(\Omega)$.

Since any function in $H^\infty(\Omega)$ can be boundedly pointwise approximated on Ω by functions in $R(\Omega)$, we see that the fully invariant subspaces of $H^p(\Omega)$ are exactly those closed subspaces which are invariant under multiplication by any function in $H^\infty(\Omega)$.

If θ is an inner function in Ω, we denote by $\theta H^p(\Omega)$ the space of all $f \in H^p(\Omega)$ which are multiples of θ. These are clearly fully invariant subspaces of $H^p(\Omega)$. The next theorem ([20] Theorem 1), which summarizes the results by Sarason, Hasumi and Voichick, states that these are the only fully invariant subspaces.

THEOREM 2.2.14. *Let $1 \leq p \leq \infty$ and let $\mathcal{M} \subseteq H^p(\Omega)$ be a fully invariant subspace of $H^p(\Omega)$. Then there exists an inner function θ such that $\mathcal{M} = \theta H^p(\Omega)$.*

Of course two inner functions θ and θ' generate the same subspace if and only if θ/θ' and θ'/θ belong to $H^p(\Omega)$ and thus to $H^\infty(\Omega)$, i.e., if and only if $\theta \equiv \theta'$.

2.3. Arithmetic of inner functions in $H^\infty(\Omega)$.

DEFINITION 2.3.1. Let f and f' be two functions in $H^\infty(\Omega)$. We say that f *divides* f' (denoted $f|f'$) if f' can be written as fg for some $g \in H^\infty(\Omega)$.

If θ and θ' are inner, then $\theta|\theta'$ and $\theta'|\theta$ if and only if $\theta \equiv \theta'$. Note also that an inner function θ divides $f \in H^\infty(\Omega)$ if and only if θ divides the inner factor of f.

PROPOSITION 2.3.2. *For any two inner functions θ and θ' in $H^\infty(\Omega)$, the following assertions are equivalent:*
(i) $\theta|\theta'$;
(ii) $\theta' H^\infty(\Omega) \subseteq \theta H^\infty(\Omega)$;
(iii) $\theta' H^2(\Omega) \subseteq \theta H^2(\Omega)$;
(iv) $|\theta'(z)| \leq c|\theta(z)|$ for some $c \geq 0$ and for all $z \in \Omega$.

DEFINITION 2.3.3. Let $\mathcal{F} = \{f_i : i \in I\}$ be a family of functions in $H^\infty(\Omega)$. An inner function θ is called the *greatest common inner divisor* of $\mathcal{F}$ if θ divides every element in $\mathcal{F}$, and if θ is a multiple of any other common inner divisor of $\mathcal{F}$. The greatest common inner divisor is denoted by $\bigwedge_{i\in I} f_i$, or by $\wedge\mathcal{F}$.

Let $\mathcal{G} = \{f_i : i \in I\}$ be a family of inner functions in $H^\infty(\Omega)$. An inner function θ is called the *least common inner multiple* of $\mathcal{G}$ if every element of $\mathcal{G}$ divides θ, and θ divides any other common inner multiple of $\mathcal{G}$. The least common inner multiple is denoted by $\bigvee_{i\in I} f_i$, or by $\vee\mathcal{G}$.

We note that the functions $\wedge\mathcal{F}$ and $\vee\mathcal{G}$ (if they exist) are not unique. In fact, two inner functions θ and θ' are greatest common inner divisors (or least common inner multiples) of the set $\mathcal{S}$, if and only if one of them is, and if $\theta \equiv \theta'$.

The first part of the following result shows that we can talk about the greatest common inner divisor of any non-empty family $\mathcal{F} \subseteq H^\infty(\Omega)$.

PROPOSITION 2.3.4. *(i) Let $\mathcal{F}$ be a non-empty family of functions in $H^\infty(\Omega)$. Then $\wedge\mathcal{F}$ exists.*
(ii) Let $\mathcal{G}$ be a family of inner functions. If $\mathcal{G}$ has a common inner multiple in $H^\infty(\Omega)$, then $\vee\mathcal{G}$ exists.

PROOF. (i) The subspace $\mathcal{M} = \bigvee\{fH^2(\Omega) : f \in \mathcal{F}\}$ of $H^2(\Omega)$ is invariant under multiplication by functions in $H^\infty(\Omega)$. Hence by Theorem 2.2.14, there exists an inner function θ such that $\mathcal{M} = \theta H^2(\Omega)$. Since $\theta H^2(\Omega) \supseteq fH^2(\Omega)$, $f \in \mathcal{F}$, it follows that $f = \theta g$ for some $g \in H^2(\Omega)$. A comparison of boundary values shows that in fact $g \in H^\infty(\Omega)$, and hence θ is a common divisor of $\mathcal{F}$. If θ' is another common divisor of $\mathcal{F}$, then $\theta' H^2(\Omega) \supseteq fH^2(\Omega)$, $f \in \mathcal{F}$, and thus $\theta' H^2(\Omega) \supseteq \mathcal{M} = \theta H^2(\Omega)$. Hence $\theta'|\theta$ by Proposition 2.3.2, and we conclude that $\theta \equiv \wedge\mathcal{F}$.

(ii) If $h \in H^\infty(\Omega)$ is a common inner multiple of the family $\mathcal{G}$, then it follows that $h \in \mathcal{N} = \bigcap\{gH^2(\Omega) : g \in \mathcal{G}\}$. Thus $\mathcal{N}$ is non-empty, and hence $\mathcal{N} = \varphi H^2(\Omega)$ for some inner function φ by Theorem 2.2.14. An argument similar to that in (i) shows that $\varphi \equiv \vee\mathcal{G}$. $\square$

The use of greatest common inner divisors is dictated by the fact that the weak*-closed ideals of $H^\infty(\Omega)$ are principal and generated by inner functions.

If $f, g \in H^\infty(\Omega)$ and $f|g$, then we denote, somewhat informally, by f/g or $\frac{f}{g}$ the function $h \in H^\infty(\Omega)$ determined by $f = gh$.

PROPOSITION 2.3.5. *If $\{\theta_i : i \in I\}$ is a family of inner divisors of the inner function θ, then $\theta/(\bigvee_{i\in I}\theta_i) \equiv \bigwedge_{i\in I}(\theta/\theta_i)$.*

PROOF. Since $\theta_j|\bigvee_{i\in I}\theta_i$, we conclude that $\theta/(\bigvee_{i\in I}\theta_i)$ divides θ/θ_j for each $j \in I$. If φ is an arbitrary common inner divisor of $\{\theta/\theta_i : i \in I\}$, then $\theta_i|\theta/\varphi$ for every $i \in I$ and consequently $(\bigvee_{i\in I}\theta_i)|\theta/\varphi$. This is equivalent to $\varphi|\theta/(\bigvee_{i\in I}\theta_i)$, and so $\theta/(\bigvee_{i\in I}\theta_i)$ is the greatest common inner divisor of $\{\theta/\theta_i : i \in I\}$. □

Common inner divisors and multiples are easier to understand in the context given by the canonical factorization of an inner function as a product of a singular function and a Blaschke product given in Theorem 2.2.11.

Let us recall that the set of positive finite measures on Γ has a lattice structure with respect to the relation: $\nu \le \nu' \Leftrightarrow \nu(A) \le \nu'(A)$ for every Borel subset A of Γ. We denote by $\nu \vee \nu'$ and $\nu \wedge \nu'$ the least upper bound and greatest lower bound of ν and ν' respectively. The measures ν and ν' are mutually singular if $\nu \wedge \nu' = 0$. The set of Blaschke functions can also be organized as a lattice with respect to the relation: $\mu \le \mu' \Leftrightarrow \mu(z) \le \mu'(z)$ for every $z \in \Omega$. Clearly then $(\mu \vee \mu')(z) = \max\{\mu(z), \mu'(z)\}$ and $(\mu \wedge \mu')(z) = \min\{\mu(z), \mu'(z)\}$, $z \in \Omega$.

PROPOSITION 2.3.6. *Let μ and μ' be Blaschke functions, ν and ν' singular positive measures on Γ, and assume that $\theta \equiv B_\mu S_\nu$ and $\theta' \equiv B_{\mu'} S_{\nu'}$. Then*
(i) $\theta \equiv \theta'$ if and only if $\mu = \mu'$ and $\nu = \nu'$;
(ii) $\theta|\theta'$ if and only if $\mu \le \mu'$ and $\nu \le \nu'$;
(iii) $\theta \vee \theta' \equiv B_{\mu\vee\mu'} S_{\nu\vee\nu'}$; $\theta \wedge \theta' \equiv B_{\mu\wedge\mu'} S_{\nu\wedge\nu'}$.

PROOF. (i) Obviously follows from the uniqueness of factorization of inner functions. If $\theta|\theta'$ and $\theta'/\theta \equiv B_{\mu''} S_{\nu''}$, then, by Lemma 2.2.12 and uniqueness of factorization, we have $\mu' = \mu + \mu'' \ge \mu$ and $\nu' = \nu + \nu'' \ge \nu$. The converse in (ii) is even more obvious, and (iii) clearly follows from (ii). □

REMARK 2.3.7. If θ is an inner function and if $\theta \equiv B_\mu S_\nu$ is its factorization provided by Theorem 2.2.11, we associate to it the subharmonic function u_θ defined by:

$$u_\theta(z) = -\sum_{\zeta\in\Omega} \mu(\zeta) g(z, \zeta) + \int_\Gamma \frac{\partial g}{\partial n}(\zeta, z) d\nu(\zeta).$$

By uniqueness of factorization in Theorem 2.2.11, the map $\theta \to u_\theta$ is well defined and, since $g(z, \zeta) \ge 0$ and $\frac{\partial g}{\partial n}(\zeta, z) \le 0$, we have that $u_\theta(z) \le 0$ for all $z \in \Omega$. More generally, if f is any function in $H^\infty(\Omega)$ and $f \equiv \theta F$ is the factorization provided by Theorem 2.2.4, then we have

$$\log|f(z)| = u_\theta(z) - \frac{1}{2\pi}\int_\Gamma \log|F(\zeta)| \frac{\partial g}{\partial n}(\zeta, z) ds + h(z),$$

where h is a harmonic function on Ω, continuous on $\overline{\Omega}$ and with constant values on each boundary component.

COROLLARY 2.3.8. *Let $\theta \equiv B_\mu S_\nu$ and $\theta' \equiv B_{\mu'} S_{\nu'}$ be two inner functions; then $\theta|\theta'$ if and only if $u_{\theta'} \le u_\theta$.*

PROOF. By Proposition 2.3.6, if $\theta|\theta'$ we have that $\mu \le \mu'$ and $\nu \le \nu'$, and thus that $u_{\theta'} \le u_\theta$. Conversely, suppose that $u_{\theta'} \le u_\theta$; since $|\theta'/\theta| = \exp(u_{\theta'} - u_\theta + h)$, where h is a harmonic function on Ω, continuous up to the boundary and with constant value on each component of Γ, then $|\theta'/\theta| \le \max\{|h(z)| : z \in \overline{\Omega}\}$. By Proposition 2.3.2 we conclude that $\theta|\theta'$. □

LEMMA 2.3.9. *There exists a constant $C > 0$ (depending only on Ω) with the following property; for each vector $\beta \in \mathbf{R}^m$ there exists a harmonic function h on Ω, continuous on $\overline{\Omega}$ and with constant values on each component of Γ such that $\|h\|_\infty = \max\{|h(z)| : z \in \overline{\Omega}\} \le C$ and, if α_j is the period of the conjugate differential of h about Γ_j, then $(\alpha_j - \beta_j)/2\pi$ is an integer, for $j = 1, \dots, m$.*

PROOF. Let A be the set of real harmonic functions in Ω which are continuous in $\overline{\Omega}$ and constant on each Γ_j, for $j = 0, 1, \dots m$. Let $\Lambda : A \to \mathbf{R}^m$ be defined by $\Lambda(h) = (\int_{\Gamma_1} d^*h, \dots, \int_{\Gamma_m} d^*h)$. Thus the jth coordinate of $\Lambda(h)$ is the period about Γ_j of the conjugate differential d^*h of h. It is well-known (see [1]) that Λ is surjective, and therefore there exists a positive constant γ with the following property: for every $\alpha \in \mathbf{R}^m$ there exists $h \in A$ such that $\Lambda(h) = \alpha$ and $\|h\|_\infty \le \gamma\|\alpha\|$.

Let $\beta \in \mathbf{R}^m$ be fixed. We can find a vector $\alpha \in R^m$ such that $(\alpha_j - \beta_j)/2\pi$ is an integer and $\alpha_j \in [-\pi, \pi)$ for $j = 1, \dots m$. By the first part of the proof, there exists $h \in A$ such that $\Lambda(h) = \alpha$ and $\|h\|_\infty \le \gamma\|\alpha\| \le \pi m^{1/2}\gamma$. The lemma follows setting $C = \pi m^{1/2}\gamma$. □

PROPOSITION 2.3.10. *There exists a positive constant C (depending only on Ω) with the following property; for any subharmonic function u on Ω of the form*

$$u(z) = -\sum_{\zeta\in\Omega} \mu(\zeta)g(z,\zeta) + \int_\Gamma \frac{\partial g}{\partial n}(\zeta, z)d\nu(\zeta),$$

there exist a harmonic function h on Ω, continuous on $\overline{\Omega}$ and constant on each component of Γ, and an inner function θ such that $\log|\theta| = u + h$ (and thus $u_\theta = u$), and $\|h\|_\infty \le C$.

PROOF. Let $u(z) = -\sum_{\zeta\in\Omega} \mu(\zeta)g(z,\zeta) + \int_\Gamma \frac{\partial g}{\partial n}(\zeta, z)d\nu(\zeta)$, and let ψ be the inner function in Ω defined by $\psi \equiv B_\mu S_\nu$; then $\log|\psi| = u + k$, where k is harmonic on Ω, continuous on $\overline{\Omega}$ and constant on each component of Γ. For $j = 1, \dots, m$, let β_j be the period of the conjugate differential of k about Γ_j; by the preceding lemma there exists a harmonic function h on Ω, continuous on $\overline{\Omega}$ and with constant values on each component of Γ such that $\|h\|_\infty \le C$ and, if α_j is the period of the conjugate differential of h about Γ_j, then $\alpha_j - \beta_j$ is an integer multiple of 2π. We conclude that the function θ defined by $\theta = \psi\exp((h - k) + i^*(h - k))$, where $^*(h - k)$ is the harmonic conjugate of $(h - k)$ in Ω, is single-valued and holomorphic on Ω. Clearly θ is inner and $\log|\theta| = u + h$. □

PROPOSITION 2.3.11. *Let $\{\theta_i : i \in I\}$ be a family of non-trivial inner divisors of the inner function θ. If $\theta_i \wedge \theta_j \equiv \mathbf{1}$ for $i, j \in I, i \ne j$, then I is at most countable.*

PROOF. Let $\theta \equiv B_\mu S_\nu$ and $\theta_i \equiv B_{\mu_i} S_{\nu_i}$, $i \in I$. It clearly suffices to prove that the two families $B = \{B_{\mu_i} : B_{\mu_i} \not\equiv \mathbf{1}\}_{i \in I}$ and $S = \{S_{\nu_i} : S_{\nu_i} \not\equiv \mathbf{1}\}_{i \in I}$, are at most countable. Let us first consider the family B. Let $i, j \in I$, $i \neq j$; since $\theta_i | \theta$ and $\theta_i \wedge \theta_j \equiv \mathbf{1}$, then by Proposition 2.3.6 we have $\mu_i \leq \mu$ and $\mu_i \wedge \mu_j = 0$, i.e.,

$$\mu_i(z) \leq \mu(z) \tag{1}$$

$$\min\{\mu_i(z), \mu_j(z)\} = 0 \tag{2}$$

for all z in Ω. Since $\sum_{z \in \Omega} \mu(z) \operatorname{dist}(z, \Gamma) < \infty$, it follows that $\mu(z) \neq 0$ for at most countably many $\{z_1, z_2, \dots\}$ in Ω. Hence by (1) we have that $\mu_i(z) = 0$ for all $i \in I$ and for all $z \notin \{z_1, z_2, \dots\}$. Moreover by (2), for all $n = 1, 2, \dots$, we have $\mu_i(z_n) \neq 0$ for at most one $i \in I$. Hence only at most countably many μ_i's are different from 0 and thus the family B is at most countable.

Let us consider now the family S. Let $i \in I$; since $\theta_i | \theta$, then by Proposition 2.3.6 $\nu_i \leq \nu$; hence by Radon-Nikodym Theorem there exists a measurable function f_i such that $0 \leq f_i \leq 1$ and $d\nu_i = f_i d\nu$. If $j \neq i$, $\theta_i \wedge \theta_j \equiv \mathbf{1}$ implies $\nu_i \wedge \nu_j = 0$, so that $\min\{f_i, f_j\} = 0$ a.e. $d\nu$, and thus if $A_i = \{z \in \Gamma : f_i(z) \neq 0\}$, then

$$\nu(A_i \cap A_j) = 0. \tag{3}$$

We prove now that (3), together with the fact that ν is a finite measure, implies that $\nu(A_i) = 0$ for all but at most countably many i's. To this purpose, for each $k = 1, 2, \dots$, we define $E_k = \{A_i : \nu(A_i) > 1/k\}$; then each E_k contains at most a finite number of A_i's. Indeed, suppose that there exists a countable family of A_i's in E_k, say $\{A_n\}_{n=1}^\infty$. Let $\{C_n\}_{n=1}^\infty$ be the family of pairwise disjoint measurable sets defined by the following relations: $C_1 = A_1$ and, for $n \geq 2$, $C_n = A_n - \cup_{k=1}^{n-1} A_k$. By (3) we have that $\nu(A_n) = \nu(C_n)$ for all n. Consequently

$$\infty > \nu(\cup_{n=1}^\infty C_n) = \sum_{n=1}^\infty \nu(C_n) = \sum_{n=1}^\infty \nu(A_n) > \sum_{n=1}^\infty \frac{1}{k} = \infty,$$

a contradiction. Hence $\cup_{k=1}^\infty E_k$ contains at most countably many A_i's. We conclude thus that only at most countably many A_i has ν-measure different from zero, so that only at most countably many ν_i are different from the zero measure. □

We conclude this section with some analytic properties of inner functions. Namely, we want to characterize the subset in Γ across which an inner function can be analytically continued.

LEMMA 2.3.12. *Let $f \not\equiv 0$ be a holomorphic function on Ω whose zeros do not accumulate at $z_0 \in \Gamma$. Then $\log|f| = u$ can be harmonically continued at z_0 if and only if f can be analytically continued at z_0.*

PROOF. Let $r > 0$ be fixed such that $\log|f|$ can be harmonically continued on $D(z_0, r) = \{z : |z - z_0| < r\}$. Let v be the harmonic conjugate of $\log|f|$ in $D(z_0, r)$. Then $g = \exp(\log|f| + iv)$ is an analytic function in $D(z_0, r)$ such that $|g(z)| = |f(z)|$, for all $z \in D(z_0, r)$. By the Maximum Modulus Principle there exists a unimodular constant γ such that $f = \gamma g$ in $D(z_0, r)$. Hence f can be analytically continued at z_0. The converse is obvious. □

LEMMA 2.3.13. *Let $\{a_n\}_{n=1}^{\infty}$ be a sequence in Ω accumulating only at the boundary Γ. Let $Z = (\{a_n\}_{n=1}^{\infty})^-$ and $z_0 \in \Gamma - Z$. Then there exists a neighborhood $\mathcal{U}$ of z_0 on which $g(z, a_n)$ can be continued harmonically for all n. Moreover, if $\sum_{n=1}^{\infty} g(z, a_n)$ converges uniformly on compact subsets of $\Omega - Z$, then $\sum_{n=1}^{\infty} g(z, a_n)$ converges uniformly on compact subsets of $(\Omega - Z) \cup \mathcal{U}$.*

PROOF. According to Koebe's theorem [15] there exists a bijective function, continuous up to the boundary, mapping Ω conformally onto a circular region (i.e., a region whose boundary curves are circles). By the invariance of Green's function, we can assume that Ω is circular. Since Γ is analytic, by means of the reflection principle, $g(z, a_n)$ can be continued across Γ harmonically, and if z^* is symmetric to z relative to Γ, then $g(z^*, a_n) = -g(z, a_n)$. So, for all $n \in \mathbf{N}$, $g(a_n, z)$ is defined in a neighborhood of Ω; more precisely, for each $\zeta \in \Gamma$, if $r = r(\zeta) > 0$ is fixed such that $D(\zeta, r)$ intersects Γ only at the component containing ζ, then $g(a_n, z)$ is defined in $\{z : |z - \zeta| < \frac{|\zeta - a_n|}{2}\} \cap D(\zeta, r)$. If $d =$ dist$(z_0, Z)/2$, then $g(a_n, z)$ is extendable for all n on $\mathcal{U} = D(z_0, d) \cap D(z_0, r)$ and clearly, if $\sum_{n=1}^{\infty} g(z, a_n)$ converges uniformly on compact subsets of $\Omega - Z$, then $\sum_{n=1}^{\infty} g(z, a_n)$ converges uniformly on compact subsets of $(\Omega - Z) \cup \mathcal{U}$. □

PROPOSITION 2.3.14. *Let B be a Blaschke product with zeros $\{a_n\}_{n=1}^{\infty}$ repeated by multiplicity. Then B can be continued analytically across an arc $\gamma \subseteq \Gamma$ if and only if $\gamma \cap (\{a_n\}_{n=1}^{\infty})^- = \emptyset$.*

PROOF. It is clear that B cannot be continued from the interior of the region to any accumulation point of $\{a_n\}_{n=1}^{\infty}$, for the extended value of B would have to be zero, while the nontangential limits of B are not. Conversely, if $\gamma \cap (\{a_n\}_{n=1}^{\infty})^- = \emptyset$, then, by Lemma 2.3.13, $\log |B|$ can be continued harmonically across γ. The result follows then from Lemma 2.3.12. □

We recall that if ν is a measure on Γ, the *support* of ν (supp(ν)) is defined to be the complement of the union of all open subsets of Γ which have ν-measure zero.

PROPOSITION 2.3.15. *Let S_ν be a singular function in $H^\infty(\Omega)$; S_ν can be analytically continued across an arc $\gamma \subseteq \Gamma$ if and only if $\gamma \cap$ supp$(\nu) = \emptyset$.*

PROOF. Let $z \in \Omega$; according to Definition 2.2.6 we have

$$u(z) = \log |S_\nu(z)| = \int_\Gamma \frac{\partial g}{\partial n}(\zeta, z) d\nu(\zeta) + h(z).$$

From Lemma 2.3.12 it is enough to prove that u can be continued harmonically at z_0 if and only if $z_0 \notin$ supp(ν). It is clear that if $z_0 \in \Gamma - \{$ supp$(\nu)\}$, then u can be continued harmonically at z_0. Actually this has nothing to do with the fact that ν is singular; indeed, the analitycity of Γ implies that $\frac{\partial g}{\partial n}(\zeta, z)$ can be extended harmonically to a neighborhood of Γ. Conversely, suppose that u extends harmonically at z_0. Then it follows by Theorem 1.3 in [18], that $z_0 \notin$ supp(ν). □

DEFINITION 2.3.16. The *support* of an inner function θ ($\text{supp}(\theta)$) consists of those points $z \in \overline{\Omega}$ such that either (i) $z \in \Omega$ and $\theta(z) = 0$; or (ii) $z \in \Gamma$ and θ cannot be continued analytically across z.

From the previous results, we clearly have $\text{supp}(\theta) = \text{supp}(B_\mu) \cup \text{supp}(S_\nu)$, if $\theta \equiv B_\mu S_\nu$. Moreover, it follows from Proposition 2.3.6 that $\text{supp}(\theta) \subseteq \text{supp}(\theta')$ whenever $\theta | \theta'$.

An important property of the support is that it can be localized in the sense of the following definition.

DEFINITION 2.3.17. Let θ be an inner function and $A \subseteq \mathbf{C}$. An inner divisor θ' of θ is a *localization* of θ to A if $\text{supp}(\theta') \subseteq \bar{A}$, $\text{supp}(\theta/\theta') \subseteq (\mathbf{C} - A)^-$, and $\theta' \wedge (\theta/\theta') \equiv \mathbf{1}$.

PROPOSITION 2.3.18. *For every inner function $\theta \equiv B_\mu S_\nu$ and every subset A of the complex plane $\mathbf{C}$, there exists a localization of θ to A.*

PROOF. It follows from the discussion above that we have only to indicate how to construct localizations of Blaschke products and singular functions. A localization of B_μ to A is the Blaschke product determined by the multiplicity function μ', defined by $\mu'(z) = 0$ if $z \notin A$, and $\mu'(z) = \mu(z)$ if $z \in A$. Analogously, a localization of S_ν to A is the singular function $S_{\nu'}$, where $\nu' = \chi_{\bar{A}} \nu$ (i.e. $\nu'(B) = \nu(B \cap \bar{A})$). $\square$

REMARK 2.3.19. We note that if θ' is a localization of θ to A, then the relation $\text{int}(A) \cap \text{supp}(\theta) \neq \emptyset$ implies that $\theta' \not\equiv \mathbf{1}$. Indeed if $\theta' \equiv \mathbf{1}$, then $\text{supp}(\theta) = \text{supp}(\theta/\theta') \subseteq \text{supp}(\theta) - \text{int}(A) \neq \text{supp}(\theta)$, a contradiction.

REMARK 2.3.20. If $\{\theta_i : i \in I\}$ is a family of divisors of θ and each θ_i is a localization of θ to A, then $\bigvee_{i \in I} \theta_i$ is a localization of θ to A. It follows that there exists a largest (in divisibility) localization of θ to A. We will call this localization the *maximal localization* of θ to A. If θ_0 is the maximal localization of θ to A, $\varphi | \theta$ and $\text{supp}(\varphi) \subseteq \overline{A}$, then $\varphi | \theta_0$.

3. The Class C_0

3.1. Construction of the functional calculus. Let H be a Hilbert space and $K \subset \mathbf{C}$ be a compact subset of the complex plane.

DEFINITION 3.1.1. If $T \in \mathcal{L}(H)$ and $\sigma(T) \subseteq K$, we say that K is a *spectral set* for the operator T if

$$\|r(T)\| \leq \max\{|r(z)| : z \in K\},$$

whenever r is a rational function with poles off K.

Note that $r(T)$ is well defined as the quotient of polynomials. If K is a spectral set for T, then the rational functional calculus $r \to r(T)$ has a unique continuous extension to $\mathrm{Rat}(\overline{\Omega})$ (cf. [11] p.389).

Hypothesis (h). From now on we assume that $T \in \mathcal{L}(H)$ is an operator with $\overline{\Omega}$ as spectral set and with no normal summand with spectrum in Γ, i.e., we assume that T has no reducing subspace $\mathcal{M} \subseteq H$ such that $T_{|\mathcal{M}}$ is normal and $\sigma(T_{|\mathcal{M}}) \subseteq \Gamma$. We say that an operator satisfying these requirements satisfies *hypothesis (h)*. This is the extension to more general regions of the notion of completely nonunitary operator (Definition 2.1.1).

The following result follows from Proposition 5.1 in [11].

LEMMA 3.1.2. *Let T be an operator satisfying hypothesis (h) and let $f \in$ Rat($\overline{\Omega}$). Then the operator $f(T)$ has $f(\overline{\Omega})$ as spectral set. Moreover for all g in Rat($f(\overline{\Omega})$) we have $(g \circ f)(T) = g(f(T))$.*

The regions Ω we are considering are conformally equivalent to circular regions by Koebe's Theorem [15]. If Λ is a conformal map of Ω onto a circular region Ω', then Λ extends to a function in $\mathrm{Rat}(\overline{\Omega})$. By the previous lemma $\Lambda(T)$ has $\Lambda(\overline{\Omega})$ as spectral set and it is easy to verify that $\Lambda(T)$ satisfies hypothesis (h) relative to $\Lambda(\Omega)$. Therefore from now on we will assume that Ω is a circular region with

$$\Gamma_0 = \mathbf{T} \qquad \text{and} \qquad \Gamma_j = \{z \in \mathbf{C} : |z - z_j| = r_j\},$$

where $|z_j| + r_j < 1$ for $j = 1, \ldots, m$, and $|z_i - z_j| > r_i + r_j$ for $i, j = 1, \ldots, m$, $i \neq j$. We set for $j = 1, \ldots, m$, $\Omega_j = \mathbf{C} - \{z \in \mathbf{C} : |z - z_j| \leq r_j\}$, and we denote by $H_0^\infty(\Omega_j)$ the closed subalgebra of $H^\infty(\Omega_j)$ consisting of those functions that vanish at ∞.

There is a natural isometric embedding of each of the spaces H^∞ and $H_0^\infty(\Omega_j)$, for $j = 1, \ldots, m$, into $H^\infty(\Omega)$ obtained simply by restricting a function f to the domain Ω. The isometric character of these embeddings follows immediately from the Maximum Modulus Principle. Moreover a sequence $\{f_n\}_{n=1}^\infty$ converges pointwise boundedly to 0 in H^∞, or in $H_0^\infty(\Omega_j)$ for $j = 1, \ldots, m$, if and only if the restricted sequence $\{f_{n|\Omega}\}_{n=1}^\infty$ converges pointwise boundedly to 0 in $H^\infty(\Omega)$ ([15] Theorem 2). Thus these embeddings are weak*-homeomorphisms ([11] Proposition 2.3), and we can consider H^∞ and $H_0^\infty(\Omega_j)$, for $j = 1, \ldots, m$, as closed subspaces of $H^\infty(\Omega)$. The following result is Theorem 7.1 in [11].

THEOREM 3.1.3. *There exist norm-continuous and weak*-continuous projections (that is idempotents) $P_j : H^\infty(\Omega) \to H^\infty(\Omega)$, $j = 0, \ldots, m$, such that*
(i) $\sum_{j=0}^m P_j = 1$,
(ii) the range of P_0 is H^∞,
(iii) for $j = 1, \ldots, m$, the range of P_j is $H_0^\infty(\Omega_j)$.

The above theorem essentially says that there is a norm-bicontinuous linear isomorphism Ψ between $H^\infty(\Omega)$ and the direct sum

$$H^\infty \oplus H_0^\infty(\Omega_1) \oplus \cdots \oplus H_0^\infty(\Omega_m),$$

which is also a weak*-homeomorphism.

THEOREM 3.1.4. *Let $T \in \mathcal{L}(H)$ be an operator satisfying hypothesis (h). Then there exists a unique norm continuous representation Φ of $H^\infty(\Omega)$ into $\mathcal{L}(H)$ such that:*
(i) $\Phi(\mathbf{1}) = \mathbf{1}_H$, where $\mathbf{1}_H \in \mathcal{L}(H)$ is the identity operator;
(ii) $\Phi(g) = T$, where $g(z) = z$ for all $z \in \Omega$;
(iii) Φ is continuous when $H^\infty(\Omega)$ and $\mathcal{L}(H)$ are given the weak-topology.*
Moreover Φ is contractive, i.e., $\|\Phi(f)\| \leq \|f\|$ for all $f \in H^\infty(\Omega)$.

PROOF. Let g_j be defined by $g_j(z) = r_j(z - z_j)^{-1}$ for $j = 1, \ldots, m$. For an operator T having $\overline{\Omega}$ as spectral set, the condition that T have no normal summand whose spectrum is contained in Γ is equivalent to the condition that the operators T and $r_j(T - z_j)^{-1} = g_j(T)$, $j = 1, \ldots, m$, are completely nonunitary.
Let f be a function in $H^\infty(\Omega)$. If $f \in H^\infty$, we simply define $\Phi(f) = f(T)$ by means of the functional calculus introduced in Proposition 2.1.4. We turn now to the case in which $f \in H_0^\infty(\Omega_j)$ for some $j \in \{1, \ldots, m\}$, and we show that, roughly speaking, the problem can be reduced to the case of the disk. To this goal we note that, since g_j^{-1} is a conformal mapping of D onto $\Omega_j \cup \{\infty\}$, the map $f \to f \circ g_j^{-1}$ is an isometric Banach algebra isomorphism of $H_0^\infty(\Omega_j)$ into H^∞, that is also a weak*-homeomorphism. We define

$$\Phi(f) = (f \circ g_j^{-1})(g_j(T)),$$

where again on the right-hand side we refer to the functional calculus of Proposition 2.1.4. In the general case in which $f \in H^\infty(\Omega)$, we know by above remarks that f can be uniquely written as $f = f_0 + f_1 + \cdots + f_m$, with $f_0 \in H^\infty$ and $f_j \in H_0^\infty(\Omega_j)$ for $j = 1, \ldots, m$. We thus define

$$\Phi(f) = \Phi(f_0) + \Phi(f_1) + \cdots + \Phi(f_m).$$

By the properties of the functional calculus for completely nonunitary operators, we immediately have that Φ is a linear map defined on $H^\infty(\Omega)$ satisfying (i), (ii) and (iii). Let A denote the subspace of $H^\infty(\Omega)$ consisting of those rational functions whose poles belong to $\{\infty, z_1, \ldots, z_m\}$. Since $\Phi(g_j) = g_j(T)$, $\Phi(g) = T$, and A is by Runge's Theorem sequentially weak*-dense in $R(\Omega)$, then $\Phi(r) = r(T)$, for all r in $R(\Omega)$.
To show that Φ is a representation we must prove that if $f, g \in H^\infty(\Omega)$, then $\Phi(fg) = \Phi(f)\Phi(g)$. This is easily established using the fact that $\Phi_{|R(\Omega)}$ is a representation, and $R(\Omega)$ is sequentially weak*-dense in $H^\infty(\Omega)$.

The uniqueness of a representation satisfying (i), (ii), and (iii) follows from the following facts: Φ is uniquely determined on A, A is sequentially weak*-dense in $R(\Omega)$ and $R(\Omega)$ is sequentially weak*-dense in $H^\infty(\Omega)$.

To prove that Φ is contractive, notice that by Proposition 2.2.2, if $f \in H^\infty(\Omega)$, then there exists a sequence $\{r_n\}_{n=1}^\infty$ in $R(\Omega)$ weak*-convergent to f, with the property that $\max\{|r_n(z)| : z \in \overline{\Omega}\} \leq \|f\|_\infty$ for every n. Then, by (iii), $\{\Phi(r_n)\}_{n=1}^\infty$ is weak*-convergent to $\Phi(f)$, and thus

$$\|\Phi(f)\| \leq \liminf_n \|\Phi(r_n)\| = \liminf_n \|r_n(T)\| \leq \liminf_n \big(\max_{z\in\overline{\Omega}} |r_n(z)|\big) \leq \|f\|_\infty,$$

where in the second inequality we used that $\overline{\Omega}$ is a spectral set for T. □

From now on we will indicate $\Phi(f)$ by $f(T)$ for all $f \in H^\infty(\Omega)$.

REMARK 3.1.5. There is a composition theorem for the functional calculus developed in the last proposition, that goes as follows. Let Λ be a conformal map of Ω onto Ω'. Then we have $(f \circ \Lambda)(T) = f(\Lambda(T))$, for all $f \in H^\infty(\Omega')$. Indeed, this is true for $r \in R(\Omega')$ by Lemma 3.1.2. The general case follows from the fact that $R(\Omega')$ is sequentially weak*-dense in $H^\infty(\Omega')$.

Set $\Omega^* = \{\bar{z} : z \in \Omega\}$, and for $f \in H^\infty(\Omega)$, let $\tilde{f}$ be the function defined on Ω^* by $\tilde{f}(z) = \overline{f(\bar{z})}$. Then $\tilde{f} \in H^\infty(\Omega^*)$ and the map $f \to \tilde{f}$ is an isometric conjugate-linear map from $H^\infty(\Omega)$ onto $H^\infty(\Omega^*)$; moreover $\tilde{f}$ is inner relative to Ω^* whenever f is inner relative to Ω.

If $T \in \mathcal{L}(H)$ satisfies hypothesis (h), then it is easy to check that the adjoint T^* satisfies hypothesis (h) relative to Ω^*.

If Φ is a representation of $H^\infty(\Omega)$ into $\mathcal{L}(H)$, associated with Φ there is an "adjoint" representation $\tilde{\Phi}$ of $H^\infty(\Omega^*)$, defined by $\tilde{\Phi}(\tilde{f}) = \Phi(f)^*$. The following proposition shows that the adjoint representation of the functional calculus of T is the functional calculus of T^*.

PROPOSITION 3.1.6. *For every $f \in H^\infty(\Omega)$ we have that $\tilde{f}(T^*) = f(T)^*$.*

PROOF. If we decompose $f \in H^\infty(\Omega)$ as $f = f_0 + \cdots + f_m$, according to Theorem 3.1.3, we have $\tilde{f} = \tilde{f}_0 + \cdots + \tilde{f}_m$, and thus $\tilde{f}(T^*) = \tilde{f}_0(T^*) + \cdots + \tilde{f}_m(T^*)$. Set $u_j = \tilde{f}_j \circ (\tilde{g}_j)^{-1} \in H^\infty$, for $j = 1, \ldots, m$; then

$$\tilde{f}_j(T^*) = u_j(\tilde{g}_j(T^*)) = (\tilde{u}_j(\tilde{g}_j(T^*)^*))^*,$$

where in the first equality we used the definition of functional calculus and in the second equality we used Proposition 2.1.5. But equality $\tilde{g}_j(T^*)^* = g_j(T)$ implies that $\tilde{f}_j(T^*) = (\tilde{u}_j(g_j(T)))^*$, and since $\tilde{u}_j = f_j \circ g_j^{-1} \in H^\infty$, then we get

$$\tilde{u}_j(g_j(T)) = (f_j \circ g_j^{-1})(g_j(T)) = f_j(T).$$

Hence $\tilde{f}_j(T^*) = f_j(T)^*$ for $j = 1, \ldots, m$; and since by Proposition 2.1.5 we also have $\tilde{f}_0(T^*) = f_0(T)^*$, the proposition is proved. □

DEFINITION 3.1.7. An operator T satisfying hypothesis (h) is said to be of *class C_0* if there exists $u \in H^\infty(\Omega) - \{0\}$ such that $u(T) = 0$; T is said to be *locally of class C_0* if for every $x \in H$ there exists $u_x \in H^\infty(\Omega) - \{0\}$ such that $u_x(T)x = 0$.

Clearly if T is of class C_0, then T is locally of class C_0; quite interestingly, the converse of this statement also holds, as we will see later.

The subspaces $\{u \in H^\infty(\Omega) : u(T) = 0\}$ and $\{u \in H^\infty(\Omega) : u(T)x = 0\}$ are fully invariant subspaces of $H^\infty(\Omega)$; hence, by Theorem 2.2.14, each one of them has the form $\theta H^\infty(\Omega)$ for some inner function θ.

DEFINITION 3.1.8. If T is of class C_0, the inner function θ such that $\theta H^\infty(\Omega) = \{u \in H^\infty(\Omega) : u(T) = 0\}$, is called the *minimal function* of T and is denoted by m_T. Analogously, if T is locally of class C_0 and $x \in H$, we denote by m_x the inner function defined by $m_x H^\infty(\Omega) = \{u \in H^\infty(\Omega) : u(T)x = 0\}$.

Note once more that the minimal function is defined to be an equivalence class of inner functions, according to Definition 2.2.5. It is convenient to allow the operator $T = 0$ on the trivial space $\{0\}$ as belonging to the class C_0; its minimal function is the function identically one.

The following result is an immediate consequence of Proposition 3.1.6.

PROPOSITION 3.1.9. *T is of class C_0 if and only if T^* is of class C_0. In this case $m_{T^*} \equiv \tilde{m}_T$.*

It seems appropriate to give some examples of operators of class C_0. Let S denote the operator in $\mathcal{L}(H^2(\Omega))$ defined by $(Sf)(z) = zf(z)$ for $z \in \Omega$, and let $\theta \in H^\infty(\Omega)$ be an inner function. We set $\mathcal{H}(\theta) = H^2(\Omega) \ominus \theta H^2(\Omega)$ and denote by $S(\theta)$ the compression of S to $\mathcal{H}(\theta)$, i.e., $S(\theta) = P_{\mathcal{H}(\theta)} S_{|\mathcal{H}(\theta)}$, where $P_{\mathcal{H}(\theta)}$ denotes the orthogonal projection onto $\mathcal{H}(\theta)$.

PROPOSITION 3.1.10. *The operator $S(\theta)$ is of class C_0 and $m_{S(\theta)} \equiv \theta$.*

PROOF. Let $N \in \mathcal{L}(L^2(\Gamma, \omega))$ be the operator of multiplication by z; then $S = N_{|H^2(\Omega)}$. By Theorem 1 in [31], the reducing subspaces for N are those of the form $\chi_E L^2(\Gamma, \omega)$ where χ_E is the characteristic function of a measurable set $E \subseteq \Gamma$. Since the boundary values of a nonzero function in $H^2(\Omega)$ cannot vanish on a set of positive measure, it follows that S does not have any normal summand. Moreover, since $\sigma(S) = \overline{\Omega}$ ([1] Theorem 2.14) and since for all $r \in R(\Omega)$ we have $r(S) = r(N)_{|H^2(\Omega)}$, then $\overline{\Omega}$ is a spectral set for S. Hence S satisfies hypothesis (h).

For each inner function θ we also have $\sigma(S(\theta)) \subseteq \overline{\Omega}$. Indeed, let $\lambda_0 \notin \overline{\Omega}$; then the matrix representing the operator $S - \lambda_0$ with respect to the decomposition $H^2(\Omega) = \theta H^2(\Omega) \oplus \mathcal{H}(\theta)$ is

$$\begin{pmatrix} (S-\lambda_0)_{|\theta H^2(\Omega)} & * \\ 0 & S(\theta) - \lambda_0 \end{pmatrix}.$$

Since $(S - \lambda_0)_{|\theta H^2(\Omega)}$ and $S - \lambda_0$ are similar, $(S - \lambda_0)_{|\theta H^2(\Omega)}$ is invertible, and thus $S(\theta) - \lambda_0$ must be invertible too.

For all $f \in R(\Omega)$ we have $f(S(\theta)) = P_{\mathcal{H}(\theta)} f(N)_{|\mathcal{H}(\theta)} = P_{\mathcal{H}(\theta)} f(S)_{|\mathcal{H}(\theta)}$, therefore $\overline{\Omega}$ is a spectral set for $S(\theta)$. Moreover the fact that S satisfies hypothesis (h) implies that the operators S and $g_j(S)$ for $j = 1, \ldots, m$, are completely nonunitary, and thus $S(\theta)$ and $g_j(S(\theta))$ for $j = 1, \ldots, m$, are completely nonunitary ([27] p.72). We thus conclude that $S(\theta)$ satisfies hypothesis (h).

We have $\theta(S(\theta)) = P_{\mathcal{H}(\theta)}\theta(S)_{|\mathcal{H}(\theta)} = 0$, since $\theta(S)H^2(\Omega) = \theta H^2(\Omega)$ is orthogonal to $\mathcal{H}(\theta)$, hence $S(\theta)$ is an operator of class C_0.

Let $J = \{u \in H^\infty(\Omega) : u(S(\theta)) = 0\}$; we have $\theta u(S(\theta)) = \theta(S(\theta))u(S(\theta)) = 0$, for all $u \in H^\infty(\Omega)$. Thus $\theta H^\infty(\Omega) \subseteq J$. Conversely, let $u \in H^\infty(\Omega)$ be such that $u(S(\theta)) = 0$. Then $P_{\mathcal{H}(\theta)}u(S)_{|\mathcal{H}(\theta)} = 0$ implies $u(S)(\mathcal{H}(\theta)) = u\mathcal{H}(\theta) \subseteq \theta H^2(\Omega)$. Hence $uH^2(\Omega) = u\mathcal{H}(\theta) + u\theta H^2(\Omega) \subseteq \theta H^2(\Omega)$, and u can be written as θg for some $g \in H^2(\Omega)$. Clearly $g \in H^\infty(\Omega)$; thus $u \in \theta H^\infty(\Omega)$, and $\{u \in H^\infty(\Omega) : u(S(\theta)) = 0\} = \theta H^2(\Omega)$. This proves that $m_{S(\theta)} \equiv \theta$. $\square$

3.2. The class C_{00}.

DEFINITION 3.2.1. An operator $T \in \mathcal{L}(H)$ satisfying hypothesis (h) is of *class* C_{0*} if the functional calculus mapping $H^\infty(\Omega) \to \mathcal{L}(H)$ is continuous when $H^\infty(\Omega)$ and $\mathcal{L}(H)$ are given respectively the weak*-topology and the strong operator topology; it is of class C_{*0} if T^* is of class C_{0*} and it is of class C_{00} if it is both of class C_{0*} and of class C_{*0} (i.e. the functional calculus mapping is continuous when $\mathcal{L}(H)$ is given the double-strong operator topology).

If we need to specify the region Ω with respect to which T is of class C_{0*} (C_{*0} or C_{00}) we will say that T is of class C_{0*} (C_{*0} or C_{00}) relative to Ω. We recall that $g_j(z) = r_j(z - z_j)^{-1}$, for $j = 1, \dots, m$.

The following result ([11] Theorem 7.2) implies that Definition 3.2.1 is consistent with Definition 2.1.8.

THEOREM 3.2.2. *Let T be an operator satisfying hypothesis (h), then the following conditions are equivalent.*
(i) Each of the sequences $\{T^n\}_{n=1}^\infty$, $\{g_j(T)^n\}_{n=1}^\infty$, $j = 1, \dots, m$, converges strongly to zero in $\mathcal{L}(H)$.
(ii) If $\{f_n\}_{n=1}^\infty$ is any sequence in $H^\infty(\Omega)$ weak-convergent to zero, then the sequence $\{f_n(T)\}_{n=1}^\infty$ converges strongly to zero in $\mathcal{L}(H)$.*

The most important result of this section is the proof that every operator of class C_0 is of class C_{00}. We start with the following two theorems, from [12] and [8] respectively.

THEOREM 3.2.3. *Let $T \in \mathcal{L}(H)$ and assume that $\overline{\Omega}$ is a spectral set for T. Then there exists an invertible operator $X \in \mathcal{L}(H)$, a Hilbert space $K \supseteq H$ and a normal operator $N \in \mathcal{L}(K)$ such that $\sigma(N) \subseteq \Gamma$ and, for every $r \in R(\Omega)$, we have $X^{-1}r(T)X = P_H r(N)_{|H}$.*

The operator N is called a *normal boundary dilation* of $X^{-1}TX$. This dilation is *minimal* if N has no nonzero reducing subspaces orthogonal onto H.

THEOREM 3.2.4. *Assume that $T \in \mathcal{L}(H)$ has $\overline{\Omega}$ as a spectral set and there exists a weak*-continuous functional calculus $H^\infty(\Omega) \to \mathcal{L}(H)$ for T. If T has a minimal normal boundary dilation N, then the spectral measure of N is absolutely continuous relative to arc-length measure on Γ.*

LEMMA 3.2.5. *Let $N \in \mathcal{L}(K)$ be a normal operator with $\sigma(N) \subseteq \Gamma$ and spectral measure E_N absolutely continuous relative to arc-length. If there exists a non-zero function $f \in H^\infty(\Omega)$ such that $f(N) = 0$, then $N = 0$.*

PROOF. Suppose that there exists a non-zero $f \in H^\infty(\Omega)$ with the property that $f(N) = 0$. Hence $f(z) = 0$ a.e. for $z \in \text{supp}(E_N)$; since f is in $H^\infty(\Omega)$, if it vanishes on a subset of Γ of positive measure, then it must be the zero function. We conclude that the arc-length measure of $\text{supp}(E_N)$ is zero. This, together with the fact that E_N is absolutely continuous relative to arc-length, imply $N = 0$. $\square$

PROPOSITION 3.2.6. *Let $T \in \mathcal{L}(H)$ be an operator of class C_0 . Then T is similar to an operator $T' \in \mathcal{L}(H)$ of class C_0 having a minimal normal boundary dilation $N \in \mathcal{L}(K)$ for some Hilbert space $K \supseteq H$. Moreover, T is of class C_{0*} (C_{*0} or C_{00}) if and only if T' is of class C_{0*} (C_{*0} or C_{00}).*

PROOF. By Theorem 3.2.3, T is similar to an operator $T' = X^{-1}TX \in \mathcal{L}(H)$ having a normal boundary dilation $N \in \mathcal{L}(K)$ that we can reduce to be minimal. Clearly $\sigma(T') \subseteq \overline{\Omega}$ and $\overline{\Omega}$ is a spectral set for T'; moreover, we can define a weak*-continuous functional calculus for T' setting $f(T') = X^{-1}f(T)X$, for all $f \in H^\infty(\Omega)$. It follows from Theorem 3.2.4 that the spectral measure of N is absolutely continuous relative to arc-length.

We claim that T' does not have normal summands with spectrum in Γ. Indeed, suppose that $T' = V \oplus T''$ with $V \in L(H_V)$ normal and $\sigma(V) \subseteq \Gamma$; since the spectral measure of N is absolutely continuous relative to arc-length, then also the spectral measure E_V of V has the same property. Let $u = m_T$; since $u(T') = X^{-1}u(T)X = 0$ and $u(V) = u(T')_{|H_V}$, then $u(V) = 0$. Hence Lemma 3.2.5 implies that $V = 0$. So we conclude that T' satisfies hypothesis (h) and T' is of class C_0. $\square$

THEOREM 3.2.7. *Let $T \in \mathcal{L}(H)$ be of class C_0; then T is of class C_{0*}.*

PROOF. By Theorem 3.2.2, it suffices to show that T and $g_j(T)$ $(j = 1, \dots, m)$ are of class C_{0*} relative to D. Let $x \in H$, and let us first consider the sequence $\{T^n\}_{n=1}^\infty$. We have $\|T^n x\|^2 = (T^{*n}T^n x, x)$; since $\|x\| \geq \|T(x)\| \geq \dots \geq \|T^n x\| \geq \dots 0$, then

$$I \geq T^*T \geq \dots \geq T^{*n}T^n \geq \dots \geq 0,$$

and thus $\{T^{*n}T^n\}_{n=1}^\infty$ converges strongly to a positive operator P ([17] Lemma 5.1.4) which has the property that $T^*PT = P$. Let now $x, y \in H$ and define

$$< x, y >= (Px, y).$$

Then $< x, x >= 0$ if and only if $x \in \ker B$, where B is the operator determined by $B^*B = P$. Hence $< \cdot, \cdot >$ induces an inner product on $H/\ker B$; we denote by H' the completion of $H/\ker B$ relative to $< \cdot, \cdot >$.

If $[x]$ is the equivalence class of x in $H/\ker B$, we define $V([x]) = [Tx]$. Then V is well defined; indeed if $[x] = [y]$, we have $(P(x-y), x-y) = 0$, and therefore

$$0 = (T^*PT(x-y), x-y) = (P(T(x-y)), T(x-y)),$$

which implies that $[Tx] = [Ty]$. Moreover, since for all $x, y \in H$

$$< x, y >= (Px, y) = (T^*PTx, y) = (PTx, Ty) =< Tx, Ty >,$$

then V is an isometry on $H/\ker B$ and then it can be extended as an isometry on H'. We call $U \in L(H'')$ the minimal unitary dilation of V, where $H'' = \bigvee_{n=-\infty}^{\infty} V^n(H')$. If $X : H \to H'$ is the natural projection, the obvious relation $XT = VX$, implies that $XT = UX$, where we consider $X : H \to H''$. We note that by Theorem 2.2.1 and Proposition 2.3.1 in [27], U can be identified with the unitary part of the isometric dilation of the completely nonunitary operator T. Since the spectral measure of the unitary dilation of T is absolutely continuous relative to arc-length (Proposition 2.1.3), then the spectral measure of U has the same property. Hence $f(U)$ makes sense for all $f \in H^\infty(\Omega)$ and, from the relation $XT = UX$, we deduce that $Xf(T) = f(U)X$ for all $f \in H^\infty(\Omega)$.

Let $f = m_T$; then $0 = Xf(T) = f(U)X$, and thus $f(U)_{|\,\mathrm{ran}(X)} = 0$. Let $[x]$ be in $\mathrm{ran}(X)$; then $[x] = Xx$ and $f(U)(Xx) = 0$. Since for all $n \in \mathbf{Z}$

$$f(U)(Xx) = 0 \Leftrightarrow U^n f(U)(Xx) = 0 \Leftrightarrow f(U)U^n(Xx) = 0,$$

then $f(U)U^n(Xx) = 0$. Hence $f(U)_{|\bigvee_{n=-\infty}^{\infty} U^n X(H)} = 0$, and consequently $f(U) = 0$. By Lemma 3.2.5 it follows that $U = 0$; this implies that X must be zero, and thus that the operators B and P are the zero operator. So the sequence $\{T^{*n}T^n\}_{n=1}^{\infty}$ converges strongly to zero and the operator T is of class C_{0*} relative to D.

In order to conclude the proof we have to show that the operators $g_j(T)$ are of class C_{0*} relative to D, for $j = 1, \ldots, m$. Applying Remark 3.1.5 to $g_j(T)$ with $\Lambda = g_j^{-1}$, we obtain $(m_T \circ g_j^{-1})(g_j(T)) = m_T(g_j^{-1}(g_j(T))) = m_T(T) = 0$, which implies that $g_j(T)$ is an operator of class C_0. We can thus repeat for $g_j(T)$ the same procedure we used for T. □

The following corollary follows from the previous theorem applied to T and T^*.

COROLLARY 3.2.8. *Every operator of class C_0 is of class C_{00}.*

REMARK 3.2.9. Before closing this section we remark that it is possible to define the notion of operators of class C_0 for a more general family of operators than those satisfying hypothesis (h), more precisely for those operators having (i) $\overline{\Omega}$ as spectral set, and (ii) a weak*-continuous functional calculus $\Phi : H^\infty(\Omega) \to \mathcal{L}(H)$. An operator belonging to this family is said to be of class C_0 if $\ker \Phi \neq \{0\}$.

On the other hand, if an operator T satisfying (i) and (ii) is of class C_0, then it satisfies hypothesis (h). Indeed, if $T = N \oplus T'$ where $N \in \mathcal{L}(H_N)$ is a normal operator having spectrum in Γ, then N satisfies (i) and (ii) (we can define $f(N) = f(T)_{|H_N}$ for all $f \in H^\infty(\Omega)$) and it is of class C_0. So by Theorem 3.2.4 its spectral measure is absolutely continuous relative to arc-length, and by Lemma 3.2.5 we conclude that $N = 0$. Since our aim is the classification of operators of class C_0, we do not loose generality in considering operators satisfying hypothesis (h).

3.3. Minimal functions and maximal vectors. The principal aim of this section is the proof that a locally C_0 operator is really of class C_0; we also prove the existence of maximal vectors as defined below.

DEFINITION 3.3.1. Let $T \in \mathcal{L}(H)$ be locally of class C_0. A vector $x \in H$ is said to be T-*maximal* (or simply *maximal*) if for every $y \in H$ we have $m_y | m_x$.

Clearly $m_x(T) = 0$ if x is maximal, and hence T is of class C_0 with $m_T \equiv m_x$. Assume that $T \in \mathcal{L}(H)$ is locally of class C_0 and $\mathcal{N} \subseteq H$ is a subspace of dimension 2. Then there exists $u \in H^\infty(\Omega)$ such that $u(T)\mathcal{N} = \{0\}$. Indeed, if $\{x_1, x_2\}$ is a basis for $\mathcal{N}$, we take $u = m_{\mathcal{N}} \equiv m_{x_1} \vee m_{x_2}$. The inner function defined in this way does not depend on the particular basis; indeed, it can be alternatively characterized by $m_{\mathcal{N}} H^\infty(\Omega) = \{u \in H^\infty(\Omega) : u(T)(\mathcal{N})\} = \{0\}$.

LEMMA 3.3.2. *Let T be locally of class C_0 and let $\mathcal{N} \subseteq H$ be a subspace of dimension 2. Then the set $A = \{x \in \mathcal{N} : m_x \not\equiv m_{\mathcal{N}}\}$ is the union of an at most countable family of one-dimensional subspaces of $\mathcal{N}$.*

PROOF. Clearly $0 \in A$ and $m_{\lambda x} \equiv m_x$ whenever λ is a non-zero scalar. We conclude that A is the union of a family of one-dimensional subspaces; say $A = \cup_{i \in I} \mathbf{C} x_i$, where x_i and x_j are linearly independent whenever $i \neq j$, $i, j \in I$. Define $\theta_i = m_{\mathcal{N}} / m_{x_i}$, $i \in I$; we have $\theta_i \not\equiv \mathbf{1}$ because $x_i \in A$. If $i \neq j$, $i, j \in I$, the vectors x_i and x_j form a basis of $\mathcal{N}$. We conclude by Proposition 2.3.5 that

$$\theta_i \wedge \theta_j \equiv m_{\mathcal{N}} / (m_{x_i} \vee m_{x_j}) \equiv m_{\mathcal{N}} / m_{\mathcal{N}} = \mathbf{1}.$$

The fact that I is at most countable follows immediately now from Proposition 2.3.11. $\square$

We recall that to any inner function θ we associate a subharmonic function u_θ (Remark 2.3.7). In the remainder of this section, if T is locally of class C_0, we will denote u_{m_x} simply by u_x.

LEMMA 3.3.3. *Let T be locally of class C_0. For each $z_0 \in \Omega$ and for every negative scalar a, the set $B = \{x \in H : u_x(z_0) \geq a\}$ is closed in H.*

PROOF. Let $\{x_n\}_{n=1}^\infty \subset B$ be a sequence convergent to x. By Proposition 2.3.10, there exist $C > 0$, a sequence $\{h_n\}_{n=1}^\infty$ of harmonic functions on Ω, continuous on $\overline{\Omega}$, constant on each component of Γ and uniformly bounded by C, and a sequence $\{\theta_n\}_{n=1}^\infty$ of inner functions such that $\log|\theta_n| = (u_{x_n} + h_n)$ for all n. Clearly $\theta_n \equiv m_{x_n}$ and, since $|\theta_n(z)| = \exp(u_{x_n} + h_n) \leq \exp(h_n) \leq \exp C$, then $\{\theta_n\}_{n=1}^\infty$ are uniformly bounded. By the Vitali-Montel Theorem, we can assume, after replacing $\{x_n\}_{n=1}^\infty$ by a subsequence if necessary, that $\{\theta_n\}_{n=1}^\infty$ is weak*-convergent to $f \in H^\infty(\Omega)$. Let $f \equiv \theta F$ be the factorization provided by Theorem 2.2.4; hence

$$\log|\theta_n| = (u_{x_n} + h_n) \to \log|f| = u_\theta - \frac{1}{2\pi} \int_\Gamma \log|F(\zeta)| \frac{\partial g}{\partial n}(\zeta, \cdot) ds + k,$$

where k is harmonic on Ω, continuous on $\overline{\Omega}$ and constant on each component of Γ.

Since $\{h_n\}_{n=1}^\infty$ are uniformly bounded we can assume, passing to a subsequence if necessary, that $\{h_n\}_{n=1}^\infty$ converges uniformly on $\overline{\Omega}$ to a function h which is harmonic on Ω, continuous on $\overline{\Omega}$ and with constant values on each component of Γ. Hence $\{u_{x_n}\}_{n=1}^\infty$ converges to a function u and $u + h = u_\theta - \frac{1}{2\pi} \int_\Gamma \log|F(\zeta)| \frac{\partial g}{\partial n}(\zeta, \cdot) ds + k$. Therefore $u - u_\theta$ is harmonic on Ω, and since $(u - u_\theta) \leq 0$ on Γ, then $(u - u_\theta) \leq 0$ on Ω by the Maximum Modulus Principle. From the fact that $u_{\theta_n}(z_0) = u_{x_n}(z_0) \geq a$ for all n, it follows that $a \leq \lim_{n\to\infty} u_{x_n}(z_0) = u(z_0) \leq u_\theta(z_0)$. By continuity of functional calculus, $\{\theta_n(T)\}_{n=1}^\infty$ is weak*-convergent (and thus weakly convergent) to $f(T)$. Therefore, if $y \in H$ we have

$$|(f(T)x, y)| \leq |((f(T) - \theta_n(T))x, y)| + |(\theta_n(T)(x - x_n), y)|$$

$$\leq |((f(T) - \theta_n(T))x, y)| + \exp C\|y\|\|x - x_n\| \to 0$$

as $n \to \infty$, where in the first inequality we made use of the relation $\theta_n(T)x_n = 0$. Since y was arbitrary, we conclude that $f(T)x = 0$, and hence $m_x|f$. This implies that $m_x|\theta$ and thus, by Corollary 2.3.8, that $u_x(z_0) \geq u_\theta(z_0)$; since $u_\theta(z_0) \geq a$, we conclude that $x \in B$. □

The following application of Baire's category argument is the key to the principal result in this section.

LEMMA 3.3.4. *Let T be locally of class C_0. Then the set $\{y \in H : \exp u_y(z_0) = \inf_{x\in H}\{\exp u_x(z_0)\}\}$ is a dense G_δ in H for every $z_0 \in \Omega$.*

PROOF. If $a = \inf_{x\in H}\{\exp u_x(z_0)\}$, then the complement of $\{y \in H : \exp u_y(z_0) = a\}$ is $\cup_{j=1}^\infty B_j$, where $B_j = \{x \in H : \exp u_x(z_0) \geq a + 1/j\}$. The preceding lemma implies that each B_j is a closed set in H, and to finish the proof it suffices to show that each B_j has empty interior. Suppose to the contrary that B_j contains the open ball $B = \{x : \|x - x_0\| < \epsilon\}$. The definition of a implies the existence of $y \in H - B_j$. We choose such a vector y and denote by $\mathcal{N}$ the linear space generated by x_0 and y. Lemma 3.3.2 implies the existence of $w \in \mathcal{N} \cap B$ such that $m_w \equiv m_y$; in particular $u_y = u_w$, from which we infer that $\exp u_w(z_0) = \exp u_y(z_0) < a + 1/j$. On the other hand $w \in B \subset B_j$, a contradiction. □

THEOREM 3.3.5. *Let T be locally of class C_0. Then there exist T-maximal vectors and the set of T-maximal vectors is a dense G_δ in H. In particular, T is of class C_0 and $m_T \equiv m_x$ for every T-maximal vector x.*

PROOF. The intersection of countably many dense G_δ sets is still a dense G_δ set, and therefore the set

$$\mathcal{M} = \{x \in H : \exp u_x(z_n) = \inf_{y\in H}\{\exp u_y(z_n)\}, n \in \mathbf{N}\}$$

is a dense G_δ for any choice of the sequence $\{z_n\}_{n=1}^\infty \subset \Omega$. Let us assume that the sequence $\{z_n\}_{n=1}^\infty$ is dense in Ω. If $n \in \mathbf{N}$, $x \in \mathcal{M}$ and $y \in H$, we then have $\exp u_x(z_n) \leq \exp u_y(z_n)$, and this extends by continuity to $\exp u_x(z) \leq \exp u_y(z)$, $z \in \Omega$. We conclude that $u_x \leq u_y$, and thus that $m_y|m_x$. Hence every element of $\mathcal{M}$ is a T-maximal vector. The remaining assertions of the theorem are obvious. □

3.4. General properties of the class C_0. We deal now with the problem of constructing invariant subspaces for an operator, and in this way reducing the study of the operator to the study of its restrictions to these subspaces; we show that for an operator T of class C_0 an approach to this problem is offered by the factorizations of its minimal function m_T.

DEFINITION 3.4.1. Let $\mathcal{M}$ be a closed subspace of H and $T \in \mathcal{L}(H)$ with $\sigma(T) \subseteq \overline{\Omega}$. $\mathcal{M}$ is said to be $R(\Omega)$*-invariant* for T if it is invariant for $r(T)$ for all $r \in R(\Omega)$.

It is immediate that if $H = H^p(\Omega)$, then $R(\Omega)$-invariant subspaces for the operator of multiplication by z coincide with the fully invariant subspaces of Definition 2.2.13. Notice also that if T is an operator satisfying hypothesis (h) and if $\mathcal{M}$ is an $R(\Omega)$-invariant subspace for T, then $u(T)\mathcal{M} \subseteq \mathcal{M}$ for all u in $H^\infty(\Omega)$, since $R(\Omega)$ is sequentially weak*-dense in $H^\infty(\Omega)$. Besides, if Λ is a conformal map from Ω onto Ω', then a subspace is $R(\Omega)$-invariant for T if and only if it is $R(\Omega')$-invariant for $\Lambda(T)$.

If $\mathcal{M}$ is $R(\Omega)$-invariant for T, then it is easy to verify that $\sigma(T_{|\mathcal{M}}) \subseteq \overline{\Omega}$ and that $r(T_{|\mathcal{M}}) = r(T)_{|\mathcal{M}}$ for all $r \in R(\Omega)$.

PROPOSITION 3.4.2. *Let $T \in \mathcal{L}(H)$ be an operator satisfying hypothesis (h), $H' \subseteq H$ be an $R(\Omega)$-invariant subspace for T, and $H'' = H \ominus H'$. Let $\begin{pmatrix} T' & X \\ 0 & T'' \end{pmatrix}$ be the matrix of T with respect to the decomposition $H = H' \oplus H''$. Then T is of class C_0 if and only if T' and T'' are operators of class C_0. If T is of class C_0, then $m_{T'}|m_T$, $m_{T''}|m_T$, and $m_T|m_{T'}m_{T''}$.*

PROOF. First note that T' and T'' satisfy hypothesis (h). For every $u \in H^\infty(\Omega)$ we have $u(T)H' \subseteq H'$ and $u(T) = \begin{pmatrix} u(T') & * \\ 0 & u(T'') \end{pmatrix}$. If $u(T) = 0$, we conclude that $u(T') = 0$ and $u(T'') = 0$, so that T', T'' are of class C_0, $m_{T'}|m_T$ and $m_{T''}|m_T$. Conversely, assume that T' and T'' are of class C_0; if $x'' \in H''$ we have $0 = m_{T''}(T'')x'' = P_{H''}m_{T''}(T)x''$, and therefore $m_{T''}(T)x'' \in H'$. Consequently $(m_{T'}m_{T''})(T)x'' = m_{T'}(T)m_{T''}(T)x'' = m_{T'}(T')m_{T''}(T)x'' = 0$. Since $m_{T'}m_{T''}(T)_{|H'} = m_{T''}m_{T'}(T)_{|H'} = m_{T''}(T')m_{T'}(T') = 0$, then we conclude that $\ker m_{T'}m_{T''}(T) \supseteq H' \cup H''$ and this implies that $m_{T'}m_{T''}(T) = 0$. Thus T is of class C_0 and $m_T|m_{T'}m_{T''}$. □

Simple examples (e.g. $T = \begin{pmatrix} T' & 0 \\ 0 & 0 \end{pmatrix}$) show that we do not usually have $m_T \equiv m_{T'}m_{T''}$. It follows from the next proposition that the equality $m_T \equiv m_{T'}m_{T''}$ is true for certain hyperinvariant subspaces.

We recall that a subspace $H' \subseteq H$ is *hyperinvariant* for T if it is invariant for every operator in $\{T\}' = \{X \in \mathcal{L}(H) : XT = TX\}$. Clearly an hyperinvariant subspace is $R(\Omega)$-invariant. If $T \in \mathcal{L}(H)$ satisfies hypothesis (h), then for all u in $H^\infty(\Omega)$ we have that $\ker u(T)$ is a hyperinvariant subspace for T.

The following result establishes a mutual correspondence between the inner divisors of m_T and some of the $R(\Omega)$-invariant subspaces for T.

PROPOSITION 3.4.3. *Let $T \in \mathcal{L}(H)$ be of class C_0, and θ be an inner divisor of m_T. If $\begin{pmatrix} T' & X \\ 0 & T'' \end{pmatrix}$ is the matrix representing T with respect to the decomposition $H = H' \oplus H''$ with $H' = \ker\theta(T)$, then $m_{T'} \equiv \theta$ and $m_{T''} \equiv m_T/\theta$.*

PROOF. We have $\theta(T') = \theta(T)_{|\ker\theta(T)} = 0$ so that $m_{T'}|\theta$. It is also clear that $\{0\} = m_T(T)H = \theta(T)(m_T/\theta)(T)H$, and thus $(m_T/\theta)(T)H'' \subseteq (m_T/\theta)(T)H \subseteq \ker\theta(T) = H'$. Consequently, $(m_T/\theta)(T'') = P_{H''}(m_T/\theta)(T)_{|H''} = 0$. We have $m_{T'}|\theta$, $m_{T''}|(m_T/\theta)$, and, by Proposition 3.4.2, $\theta(m_T/\theta) = m_T|m_{T'}m_{T''}$. Hence $m_{T'} \equiv \theta$ and $m_{T''} \equiv m_T/\theta$. □

DEFINITION 3.4.4. An operator $X \in \mathcal{L}(H, H')$ is a *quasiaffinity* if it is one-to-one and with dense range.

DEFINITION 3.4.5. Let $T \in \mathcal{L}(H)$ be an operator satisfying hypothesis (h). Then we define the class $K_T^\infty(\Omega) = \{u \in H^\infty(\Omega) : u(T) \text{ is a quasiaffinity}\}$.

We indicate for simplicity $K_T^\infty = K_T^\infty(\Omega)$. If $u \in K_T^\infty$, then in general $u(T)^{-1}$ is a discontinuous closed and densely defined operator. The class K_T^∞ is important in introducing the functional calculus with rational functions:

$$\frac{v}{u}(T) = u(T)^{-1}v(T), \quad u \in K_T^\infty \quad \text{and} \quad v \in H^\infty(\Omega).$$

PROPOSITION 3.4.6. *For every operator T of class C_0 we have $K_T^\infty = \{u \in H^\infty(\Omega) : u \wedge m_T \equiv \mathbf{1}\}$. Moreover for $u \in H^\infty(\Omega)$ we have $\ker u(T) = \{0\}$ if and only if $\ker u(T)^* = \{0\}$, and for all $u \in H^\infty(\Omega)$ we have $\ker u(T) = \ker(u \wedge m_T)(T)$.*

PROOF. Assume that $u \wedge m_T \equiv \mathbf{1}$. If $x \in \ker u(T)$, then $m_x|u$ and $m_x|m_T$ so that $m_x \equiv \mathbf{1}$. This clearly implies $x = 0$ and therefore $\ker u(T) = \{0\}$. Analogously, $u \wedge m_T \equiv \mathbf{1}$ implies that $\tilde{u} \wedge m_{T^*} \equiv \mathbf{1}$, and thus that $\ker u(T)^* = \ker\tilde{u}(T^*) = \{0\}$. Conversely, if $u \wedge m_T \equiv \theta \not\equiv \mathbf{1}$, then $\ker u(T) \supseteq \ker\theta(T)$, and $\ker\theta(T) \neq \{0\}$ by Proposition 3.4.3. The last assertion is clear. □

We collect now the most important facts about the class K_T^∞.

(i) For all $u \in K_T^\infty$ we have $(u(T)^{-1})^* = (u(T)^*)^{-1} = (\tilde{u}(T^*))^{-1}$; hence we get that $\tilde{u} \in K_{T^*}^\infty(\Omega^*)$ and $K_{T^*}^\infty(\Omega^*) = \{\tilde{u} : u \in K_T^\infty(\Omega)\}$.

(ii) K_T^∞ is multiplicative.

(iii) If $u \in H^\infty(\Omega)$ is outer, then $u \in K_T^\infty$.

(iv) The condition in (iii) on the function u to be outer, turns out to be also necessary in the following sense: if $u \in H^\infty(\Omega)$ is not an outer function, then there exists $T \in \mathcal{L}(H)$, $H \neq \{0\}$, of class C_0, such that $u(T) = 0$. Indeed, if $u = \theta U$ is the canonical factorization of u with θ inner and U outer, since θ is non-trivial, there exists $T \in \mathcal{L}(H)$, $H \neq \{0\}$, such that $m_T \equiv \theta$; then $u(T) = \theta(T)U(T) = 0$.

(v) By (iii) and (iv) we conclude that if $u \in H^\infty(\Omega)$, then $u \in K_T^\infty$ for all T of class C_0 if and only if u is outer.

PROPOSITION 3.4.7. *Let $T \in \mathcal{L}(H)$ be an operator of class C_0. If T is not a scalar, then it has a non-trivial hyperinvariant subspace.*

PROOF. By Proposition 3.4.3, inner divisors θ of m_T are uniquely determined (up to invertible inner functions) by the hyperinvariant subspaces $\ker\theta(T)$. Thus if $\theta \not\equiv \mathbf{1}$ and $\theta \not\equiv m_T$, then $\ker\theta(T)$ is a non-trivial hyperinvariant subspace for T. If we assume that m_T has no non-trivial inner divisors, then m_T must be a Blaschke factor $m_T(z) = B(z)$, i.e., $\log|B(z)| = g(z,a) + h$, where h is a harmonic function, continuous on $\overline{\Omega}$ and with constant values on each component of Γ. Then the fact that $m_T(a) = 0$ implies that $m_T(z) = (z-a)g(z)$ for some function g in $H^\infty(\Omega)$, and thus $0 = m_T(T) = (T-aI)g(T)$. By Proposition 3.4.6, since $g \wedge m_T \equiv \mathbf{1}$, we have that $\ker g(T) = \{0\}$ and $\ker g(T)^* = \{0\}$. So $g(T)^{-1}$ exists and is densely defined. We can therefore conclude that $T - aI = 0$, which is a contradiction. □

The hyperinvariant subspaces considered above will help us to determine the spectrum $\sigma(T)$ of an operator T of class C_0 in terms of the minimal function m_T. We denote by $\sigma_p(T)$ the point spectrum of T, i.e., the set of all eigenvalues of T. A well-known fact in linear algebra is that the zeros of the minimal polinomial of a matrix are exactly the characteristic values of the matrix. An analogous fact holds for the class C_0.

THEOREM 3.4.8. *For every operator T of class C_0 we have $\sigma(T) = supp(m_T)$ and $\sigma_p(T) = supp(m_T) \cap \Omega$.*

PROOF. Let us assume that $z_0 \in \overline{\Omega} - \{\mathrm{supp}(m_T)\}$, then $m_T(z_0) \neq 0$. It follows that we have a factorization of the form $m_T(z) - m_T(z_0) = (z - z_0)g(z)$, for some $g \in H^\infty(\Omega)$. Functional calculus then gives

$$-m_T(z_0)I = m_T(T) - m_T(z_0)I = (T - z_0I)g(T) = g(T)(T - z_0I)$$

and thus $(T - z_0I)$ has an inverse, which is $-(1/m_T(z_0))g(T)$. Hence we have $\sigma(T) \subseteq \mathrm{supp}(m_T)$. Suppose now that $\sigma(T) \neq \mathrm{supp}(m_T)$; in this case we can find an open set O such that $\overline{O} \cap \sigma(T) = \emptyset$ but $O \cap \mathrm{supp}(m_T) \neq \emptyset$. Let θ be a localization of m_T to O whose existence is guaranteed by Proposition 2.3.18; from Remark 2.3.19 we have that θ is non-trivial, and hence by Proposition 3.4.3 we conclude that $\ker\theta(T) \neq \{0\}$. Let $\begin{pmatrix} T' & X \\ 0 & T'' \end{pmatrix}$ be the matrix representing T with respect to the decomposition $H = \ker\theta(T) \oplus H''$. We have $m_{T'} \equiv \theta$ and because of the first part of the proof we have $\sigma(T') \subseteq \mathrm{supp}(\theta) \subseteq \overline{O}$; on the other hand, since $\ker\theta(T)$ is a hyperinvariant subspace for T, we have $\sigma(T') \subseteq \sigma(T) \subseteq \mathbf{C} - \overline{O}$. These two relations are in contradiction, hence $\sigma(T) = \mathrm{supp}(m_T)$.

For the point spectrum note that T and $g_j(T)$, $j = 1, \ldots, m$, do not have eigenvalues of absolute value 1, since they are completely nonunitary. So T does not have eigenvalues on Γ and $\sigma_p(T) \subseteq \sigma(T) \cap \Omega \subseteq \mathrm{supp}(m_T) \cap \Omega$. For the opposite inclusion, note that each $z_0 \in \mathrm{supp}(m_T) \cap \Omega$ is a zero of m_T and thus m_T can be written as $(z - z_0)g(z)$. By the functional calculus $0 = (T - z_0I)g(T)$; if $g \in m_T H^\infty(\Omega)$ then $m_T | g$, a contradiction. Hence $g(T) \neq 0$ and thus $\ker(T - z_0I) \neq \{0\}$. □

REMARK 3.4.9. For an operator T of class C_0 any $R(\Omega)$-invariant subspace $\mathcal{M}$ is actually *rationally invariant*, i.e., it is invariant for $r(T)$ for all rational functions r with poles off $\sigma(T)$. Indeed, if $T' = T_{|\mathcal{M}}$, then, by Proposition 3.4.2,

$m_{T'}|m_T$, and thus $\operatorname{supp}(m_{T'}) \subseteq \operatorname{supp}(m_T)$. By Theorem 3.4.8 we then have $\sigma(T') \subseteq \sigma(T)$.

The following corollary follows from theorems on Blaschke products.

COROLLARY 3.4.10. *Let T be an operator of class C_0 and let $\{\lambda_0, \lambda_1, \dots\}$ be its eigenvalues in Ω. Then $\sum_{n=0}^{\infty} g(z, \lambda_n) < \infty$ uniformly on compact subsets of $\Omega - \{\lambda_n\}_{n=0}^{\infty}$.*

COROLLARY 3.4.11. *There exists T of class C_0 such that $\sigma(T) = \Gamma$.*

We conclude this section with a few remarks about the localization of the spectrum of a C_0 operator.

If $T \in \mathcal{L}(H)$, a subspace $\mathcal{M} \subseteq H$ is a *maximal spectral subspace* for T if it is invariant, and for every invariant subspace $\mathcal{N}$ of T such that $\sigma(T_{|\mathcal{N}}) \subseteq \sigma(T_{|\mathcal{M}})$, we have $\mathcal{N} \subseteq \mathcal{M}$.

T is said to be *decomposable* if for every finite open covering $\{G_1, G_2, \dots, G_n\}$ of $\sigma(T)$ there exist maximal spectral subspaces $\mathcal{M}_1, \mathcal{M}_2, \dots, \mathcal{M}_n$ for T such that $\sigma(T_{|\mathcal{M}_j}) \subset G_j$ for all j, and $H = \mathcal{M}_1 + \mathcal{M}_2 + \cdots \mathcal{M}_n$.

LEMMA 3.4.12. *Let $T \in \mathcal{L}(H)$ be an operator of class C_0, $A \subset \mathbf{C}$ be a subset of the complex plane and denote by θ the maximal localization of m_T to A. Then the space $\ker \theta(T)$ is a maximal spectral subspace for T and $\sigma(T_{|\ker \theta(T)}) \subseteq \overline{A}$.*

PROOF. We have $\sigma(T_{|\ker\theta(T)}) \subseteq \operatorname{supp}(\theta) \subseteq \overline{A}$, from Proposition 3.4.3. If $\mathcal{N}$ is an invariant subspace for T with $\sigma(T_{|\mathcal{N}}) \subseteq \sigma(T_{|\ker\theta(T)}) \subseteq \overline{\Omega}$, then $\mathcal{N}$ is $R(\Omega)$-invariant. Hence, by Proposition 3.4.2, $T_{|\mathcal{N}}$ is of class C_0 and by Theorem 3.4.8, $\sigma(T_{|\mathcal{N}}) = \operatorname{supp}(m_{T_{|\mathcal{N}}})$. So $\operatorname{supp}(m_{T_{|\mathcal{N}}}) \subseteq \operatorname{supp}(\theta) \subseteq \overline{A}$ and, by Remark 2.3.19, $m_{T_{|\mathcal{N}}}|\theta$; thus we have $\theta(T_{|\mathcal{N}}) = 0$ and therefore we conclude that $\mathcal{N} \subseteq \ker\theta(T)$, since $\theta(T)_{|\mathcal{N}} = \theta(T_{|\mathcal{N}})$. □

LEMMA 3.4.13. *Let $\{G_1, G_2, \dots, G_n\}$ be a finite open covering of $\sigma(T)$, where $T \in \mathcal{L}(H)$ is an operator of class C_0. For each $j \in \{1, \dots, n\}$, choose a localization θ_j of m_T to G_j. Then we have $\ker\theta_1(T) + \ker\theta_2(T) + \cdots + \ker\theta_n(T) = H$.*

PROOF. Let $\{D_1, D_2, \dots, D_n\}$ be an open covering of $\sigma(T)$ such that $\overline{D_j} \subseteq G_j$, for $1 \le j \le n$. Since $\operatorname{supp}(m_T/\theta_j) \cap G_j = \emptyset$, then for $1 \le j \le n$,

$$\inf\{|(m_T/\theta_j)(z)| : z \in D_j \cap \Omega\} > 0. \tag{1}$$

On the other hand we have $\inf\{|m_T(z)| : z \in \Omega - (\cup_{j=1}^n D_j)\} > 0$, since $\cup_{j=1}^n D_j$ covers $\operatorname{supp}(m_T)$. Therefore, for $1 \le j \le n$,

$$\inf\{|(m_T/\theta_j)(z)| : z \in \Omega - (\cup_{j=1}^n D_j)\} > 0. \tag{2}$$

Combining (1) and (2) we get

$$\inf\Big\{\sum_{j=1}^{n} |(m_T/\theta_j)(z)| : z \in \Omega\Big\} > 0,$$

and by Carleson's Corona Theorem ([13] Theorem 6.6.3), there exist functions $u_1, \dots, u_n \in H^\infty(\Omega)$ such that $\sum_{j=1}^n u_j m_T/\theta_j = \mathbf{1}$. Functional calculus then yields

$$\sum_{j=1}^n (m_T/\theta_j)(T)u_j(T)x = x, \qquad x \in H.$$

Since the equality $\theta_j(T)(m_T/\theta_j)(T)u_j(T) = m_T(T)u_j(T) = 0$ means that the vector $(m_T/\theta_j)(T)u_j(T)x$ belongs to $\ker\theta_j(T)$ for all $j \in \{1, \dots, n\}$, then the proof is concluded. □

PROPOSITION 3.4.14. *Every operator of class C_0 is decomposable.*

PROOF. Let $\{G_1, G_2, \dots, G_n\}$ be an open covering of $\sigma(T)$, and choose an open covering $\{D_1, D_2, \dots, D_n\}$ of $\sigma(T)$ such that $\overline{D_j} \subseteq G_j$, $1 \le j \le n$. For each j, let θ_j be the maximal localization of m_T to D_j. Then the spaces $\ker\theta_j(T)$ are maximal spectral by Lemma 3.4.12, $\sigma(T_{|\ker\theta_j(T)}) = \mathrm{supp}(\theta_j) \subseteq \overline{D_j} \subseteq G_j$, and $\ker\theta_1(T) + \ker\theta_2(T) + \cdots + \ker\theta_n(T) = H$ by the last lemma. □

4. Classification Theory

4.1. Jordan blocks.

DEFINITION 4.1.1. Given an inner function θ, the *Jordan block* $S(\theta)$ is the operator acting on the space $\mathcal{H}(\theta) = H^2(\Omega) \ominus \theta H^2(\Omega)$ as follows:

$$S(\theta) = P_{\mathcal{H}(\theta)} S_{|\mathcal{H}(\theta)},$$

where $S \in \mathcal{L}(H^2(\Omega))$ is defined by $(Sf)(z) = zf(z)$.

These operators were introduced in the first section of chapter 3, where we proved that they are operators of class C_0 and $m_{S(\theta)} \equiv \theta$ (Proposition 3.1.10).

If we assume that $\mathcal{M} \subset H^2$ is a co-invariant subspace for the unilateral shift U_+, and $T \in \mathcal{L}(\mathcal{M})$ is defined by $T = P_{\mathcal{M}} U_{+|\mathcal{M}}$, it was proved by Sarason [23] that every operator $X \in \mathcal{L}(\mathcal{M})$ commuting with T has the form $X = P_{\mathcal{M}} Y_{|\mathcal{M}}$, where Y is an analytic Toeplitz operator. Moreover, Y can be chosen so that $\|Y\| = \|X\|$. This result was extended by H. Bercovici and A. Zucchi [9] to the case when the unit disk is replaced by a finitely connected region in $\mathbf{C}$.

Let θ, θ' be two inner functions in $H^\infty(\Omega)$; it is easy to verify that every function $u \in H^\infty(\Omega)$ which satisfies $\theta' | u\theta$, determines an operator $X \in \mathcal{L}(\mathcal{H}(\theta), \mathcal{H}(\theta'))$ such that $XS(\theta) = S(\theta')X$, by the formula $X = P_{\mathcal{H}(\theta')} u(S)_{|\mathcal{H}(\theta)}$. Note that $X = 0$ if and only if $\theta' | u$.

The following theorem from [9] establishes the converse of this fact.

THEOREM 4.1.2. *There exists a constant $k \geq 1$, depending only on Ω, with the following property. Given two inner functions $\theta, \theta' \in H^\infty(\Omega)$, and an operator $X \in \mathcal{L}(\mathcal{H}(\theta), \mathcal{H}(\theta'))$ satisfying $XS(\theta) = S(\theta')X$, there exists u in $H^\infty(\Omega)$ such that:*
(i) $\|u\|_\infty \leq k\|X\|$;
(ii) $\theta' | u\theta$; and
(iii) $X = P_{\mathcal{H}(\theta')} u(S)_{|\mathcal{H}(\theta)}$.

COROLLARY 4.1.3. *There exists a constant $k \geq 1$, depending only on Ω, with the following property. Given any inner function θ and any operator $X \in \mathcal{L}(\mathcal{H}(\theta))$ commuting with $S(\theta)$, there exists $u \in H^\infty(\Omega)$ with $\|u\| \leq k\|X\|$ such that $X = u(S(\theta))$. In particular there exists a bijective bicontinuous linear map from $H^\infty(\Omega)/\theta H^\infty(\Omega)$ onto $\{S(\theta)\}'$.*

PROOF. The condition $\theta | u\theta$ is trivially satisfied for all $u \in H^\infty(\Omega)$, and thus the existence of u clearly follows from Theorem 4.1.2. The functional calculus of $S(\theta)$ determines an operator $\Lambda : H^\infty(\Omega)/\theta H^\infty(\Omega) \to \{S(\theta)\}'$ of norm ≤ 1. The existence for each $X \in \{S(\theta)\}'$ of a function u with $\|u\| \leq k\|X\|$ and $X = \Lambda(u)$, implies that Λ is bijective and $\|\Lambda^{-1}(X)\| \leq k\|X\|$ for all $X \in \{S(\theta)\}'$. □

Let us introduce some additional notation. If $T \in \mathcal{L}(H)$ satisfies hypothesis (h), then we denote by $\mathcal{A}_T$ the weak*-closed subalgebra of $\mathcal{L}(H)$ generated by T, I and $(T - z_j)^{-1}$ for $j = 1, \ldots, m$, and by $\mathcal{W}_T$ the weakly-closed subalgebra of $\mathcal{L}(H)$ generated by T, I and $(T - z_j)^{-1}$ for $j = 1, \ldots, m$. Since $\{T\}'$ is always a weakly-closed algebra, we naturally have $\mathcal{A}_T \subseteq \mathcal{W}_T \subseteq \{T\}'$.

The following proposition is a direct consequence of Theorem 4.1.2.

PROPOSITION 4.1.4. *If θ is inner, then we have $\mathcal{A}_{S(\theta)} = \mathcal{W}_{S(\theta)} = \{S(\theta)\}' = \{u(S(\theta)) : u \in H^\infty(\Omega)\}$.*

DEFINITION 4.1.5. An operator $T \in \mathcal{L}(H)$ is called a *quasiaffine* transform of an operator $T' \in \mathcal{L}(H')$ $(T \prec T')$ if there exists a quasiaffinity $X \in \mathcal{L}(H, H')$ such that $T'X = XT$. T and T' are *quasisimilar* $(T \sim T')$ if $T \prec T'$ and $T' \prec T$.

PROPOSITION 4.1.6. *Let $T \in \mathcal{L}(H)$ and $T' \in \mathcal{L}(H')$ be two operators satisfying hypothesis (h) such that $T \prec T'$. If one operator is of class C_0, then so is the other and their minimal functions coincide.*

As we said before, the minimal function of an operator of class C_0 plays a role analogous in many respects to the well-known role of the minimal polynomials of finite matrices in linear algebra. We know e.g. that similar matrices have the same minimal polynomial. The preceding proposition shows that quasisimilar operators of class C_0 have the same minimal function.

DEFINITION 4.1.7. Let $T \in \mathcal{L}(H)$ be an operator with spectrum in $\overline{\Omega}$. A subset $\mathcal{M} \subseteq H$ with the property that

$$\bigvee\{r(T)m : r \in R(\Omega), m \in \mathcal{M}\} = H,$$

is called an $R(\Omega)$*-generating* set for T. The *multiplicity* μ_T of T is the smallest cardinality of an $R(\Omega)$-generating set for T. The operator T is said to be *multiplicity-free* if $\mu_T = 1$. If $\mu_T = 1$, any vector $x \in H$ such that $\bigvee\{r(T)x : r \in R(\Omega)\} = H$ is said to be $R(\Omega)$*-cyclic* for T.

Thus μ_T is the number of $R(\Omega)$-cyclic subspaces that are needed to generate H, where an $R(\Omega)$-cyclic subspace for T is a subspace of the form $\bigvee\{r(T)x : r \in R(\Omega)\}$ for some $x \in H$. We observe that if T has an $R(\Omega)$-cyclic vector x, then, since $\{r(T) : r \in R(\Omega)\}$ is a separable subalgebra of $\mathcal{L}(H)$, we have that H is separable.

REMARK 4.1.8. Let $T \in \mathcal{L}(H)$ be an operator of class C_0. If $x \in H$ is $R(\Omega)$-cyclic for T, then x is $T-maximal$. Indeed, if x is $R(\Omega)$-cyclic for T, and if $y = r(T)x$ with $r \in R(\Omega)$, then $m_x(T)y = m_x(T)r(T)x = r(T)m_x(T)x = 0$, by definition of m_x.

LEMMA 4.1.9. *Let $T \in \mathcal{L}(H)$ and $T' \in \mathcal{L}(H')$ be operators with spectrum in $\overline{\Omega}$. If there exists an operator X such that $XT = T'X$ and $(XH)^- = H'$, then $\mu_{T'} \leq \mu_T$.*

DEFINITION 4.1.10. Let T be an operator of class C_0. We define the *multiplicity function* ν_T of T as $\nu_T(\theta) = \mu_{T_{|(\operatorname{ran}\theta(T))^-}}$, for any inner function θ.

Clearly $\mu_T \leq \dim(H) \leq \aleph_0\mu_T$, and thus $\nu_T(\theta) \leq \dim(\operatorname{ran}\theta(T))^- \leq \aleph_0\nu_T(\theta)$; in particular we have $\nu_T(\theta) = \dim(\operatorname{ran}\theta(T))^-$ in the case in which $\nu_T(\theta) \geq \aleph_0$, or $\dim(\operatorname{ran}\theta(T))^- > \aleph_0$.

PROPOSITION 4.1.11. *Let $T \in \mathcal{L}(H)$ and $T' \in \mathcal{L}(H')$ be two operators of class C_0 such that $T \prec T'$. Then for all inner functions θ we have $\nu_{T'}(\theta) \leq \nu_T(\theta)$. In particular the multiplicity function is a quasisimilarity invariant for the class C_0.*

PROOF. If X is a quasiaffinity such that $T'X = XT$, we deduce that $T_{|(\operatorname{ran}\theta(T))^-} \prec T'_{|(\operatorname{ran}\theta(T'))^-}$ via the quasiaffinity $X_{|(\operatorname{ran}\theta(T))^-}$. By Lemma 4.1.9 we have $\nu_{T'}(\theta) \leq \nu_T(\theta)$ and the proposition is proved. □

LEMMA 4.1.12. *Let $N \in \mathcal{L}(L^2(\Gamma,\omega))$ be the operator of multiplication by z and let $\mathcal{M}, \mathcal{N} \subseteq L^2(\Gamma,\omega)$ be two non-reducing $R(\Omega)$-invariant subspaces for N, such that $\mathcal{N} \subseteq \mathcal{M}$. Then the operator $T = P_{\mathcal{M}\ominus\mathcal{N}} N_{|\mathcal{M}\ominus\mathcal{N}}$ is similar to $S(\theta)$ for some inner function θ.*

PROOF. From Theorem 1 in [31], there exist invertible functions φ and ψ in $L^\infty(\Gamma,\omega)$ with $|\varphi|$ and $|\psi|$ constant a.e. on each component of Γ such that $\mathcal{M} = \psi H^2(\Omega)$ and $\mathcal{N} = \varphi H^2(\Omega)$. Since $\mathcal{M} \supseteq \mathcal{N}$ there exists $\theta \in H^2(\Omega)$ such that $\varphi = \psi\theta$, and since φ and ψ are in $L^\infty(\Gamma,\omega)$, then θ is bounded. We conclude that θ is an inner function and

$$\mathcal{M} \ominus \mathcal{N} = \psi H^2(\Omega) \ominus \psi\theta H^2(\Omega).$$

Let $Y : \mathcal{M} \ominus \mathcal{N} \to \psi H^2(\Omega)/\psi\theta H^2(\Omega)$ be the invertible operator that maps each $f \in \mathcal{M} \ominus \mathcal{N}$ into its equivalence class $[f] \in \psi H^2(\Omega)/\psi\theta H^2(\Omega)$. Clearly for $[f] \in \psi H^2(\Omega)/\psi\theta H^2(\Omega)$ we have $Y^{-1}([f]) = P_{\mathcal{M}\ominus\mathcal{N}} f$.

Similarly let $L : \mathcal{H}(\theta) \to H^2(\Omega)/\theta H^2(\Omega)$ be the invertible operator sending each $f \in \mathcal{H}(\theta)$ into its equivalence class $[f] \in H^2(\Omega)/\theta H^2(\Omega)$.

Let $M_\psi : H^2(\Omega) \to \psi H^2(\Omega)$ be the operator of multiplication by ψ. Then M_ψ induces an invertible operator M_ψ from $H^2(\Omega)/\theta H^2(\Omega)$ into $\psi H^2(\Omega)/\psi\theta H^2(\Omega)$, defined by $M_\psi([f]) = [\psi f]$. We thus have

$$M_\psi(H^2(\Omega)/\theta H^2(\Omega)) = \psi H^2(\Omega)/\psi\theta H^2(\Omega),$$

that is, $M_\psi(L(\mathcal{H}(\theta))) = Y(\mathcal{M} \ominus \mathcal{N})$.

Let $X \in \mathcal{L}(\mathcal{H}(\theta), \mathcal{M}\ominus\mathcal{N})$ be the invertible operator defined by $X = Y^{-1}M_\psi L$, and let $f \in \mathcal{H}(\theta)$. Then

$$XS(\theta)f = Y^{-1}M_\psi L(P_{\mathcal{H}(\theta)}zf) = Y^{-1}M_\psi([P_{\mathcal{H}(\theta)}zf])$$
$$= Y^{-1}M_\psi([zf]) = Y^{-1}([\psi zf]) = P_{\mathcal{M}\ominus\mathcal{N}}(\psi zf),$$

where in the third equality we used the fact that $[P_{\mathcal{H}(\theta)}zf] = [zf]$.

On the other hand we have

$$TXf = P_{\mathcal{M}\ominus\mathcal{N}}(zY^{-1}M_\psi Lf) = P_{\mathcal{M}\ominus\mathcal{N}}(zY^{-1}M_\psi([f]))$$
$$= P_{\mathcal{M}\ominus\mathcal{N}}(zY^{-1}([\psi f])) = P_{\mathcal{M}\ominus\mathcal{N}}(zP_{\mathcal{M}\ominus\mathcal{N}}\psi f) = P_{\mathcal{M}\ominus\mathcal{N}}(\psi zf);$$

indeed, if $\psi f = \psi f_1 \oplus \psi\theta f_2 \in (\mathcal{M} \ominus \mathcal{N}) \oplus \mathcal{N}$, then it follows that

$$P_{\mathcal{M}\ominus\mathcal{N}}(zP_{\mathcal{M}\ominus\mathcal{N}}\psi f) = P_{\mathcal{M}\ominus\mathcal{N}}z\psi f_1 = P_{\mathcal{M}\ominus\mathcal{N}}z\psi f.$$

We conclude that $XS(\theta) = TX$. □

THEOREM 4.1.13. *$S(\theta)^*$ is similar to $S(\tilde{\theta})$ for every inner function θ.*

PROOF. Let N be the operator of multiplication by z on $L^2(\Gamma,\omega)$, and $\tilde{N}$ be the operator of multiplication by z on $L^2(\Gamma^*,\omega^*)$, where $\Gamma^* = \partial\Omega^*$ and ω^* is the harmonic measure on Γ^* for the point $\bar{z}_0$. Let $U : L^2(\Gamma,\omega) \to L^2(\Gamma^*,\omega^*)$ be the unitary operator defined by $(Uf)(z) = f(\bar{z})$. Then clearly $UN^* = \tilde{N}U$, since $(N^*f)(z) = \bar{z}f(z)$. The relation $UN^* = \tilde{N}U$ implies that the subspaces

$$U((\theta H^2(\Omega))^\perp) = \mathcal{M} \qquad \text{and} \qquad U(H^2(\Omega)^\perp) = \mathcal{N}$$

are $R(\Omega^*)$-invariant for $\tilde{N}$, because $(\theta H^2(\Omega))^\perp$ and $H^2(\Omega)^\perp$ are $R(\Omega^*)$-invariant for N^*. Since U is unitary, we have $U(\theta H^2(\Omega)) = \mathcal{M}^\perp$ and $U(H^2(\Omega)) = \mathcal{N}^\perp$. Thus, if $f \in \mathcal{H}(\theta)$, then $Vf \in \mathcal{M} \ominus \mathcal{N}$. Hence $W = U_{|\mathcal{H}(\theta)}$ can be viewed as a unitary operator from $\mathcal{H}(\theta)$ onto $\mathcal{M} \ominus \mathcal{N}$. Set $T = P_{\mathcal{M}\ominus\mathcal{N}}\tilde{N}_{|\mathcal{M}\ominus\mathcal{N}}$; for $f \in \mathcal{H}(\theta)$ we have

$$TWf = P_{\mathcal{M}\ominus\mathcal{N}}\tilde{N}Uf = P_{\mathcal{M}\ominus\mathcal{N}}UN^*f,$$

and, since $N^*f \in (\theta H^2(\Omega))^\perp$, it follows that $UN^*f \in \mathcal{M}$, and thus

$$P_{\mathcal{M}\ominus\mathcal{N}}UN^*f = P_{\mathcal{N}^\perp}UN^*f.$$

On the other hand, since $N^*f \in (\theta H^2(\Omega))^\perp$, we have

$$WS(\theta)^*f = UP_{\mathcal{H}(\theta)}N^*f = UP_{H^2(\Omega)}N^*f,$$

hence we get $UP_{H^2(\Omega)}N^*f = P_{\mathcal{N}^\perp}UN^*f$, and thus we obtain $TW = WS(\theta)^*$.

The subspaces $\mathcal{M}$ and $\mathcal{N}$ are not reducing for $\tilde{N}$, because $(\theta H^2(\Omega))^\perp$ and $H^2(\Omega)^\perp$ are not reducing for N^*. Hence, from the preceding lemma, there exist an inner function $g \in H^\infty(\Omega^*)$, and an invertible operator $X \in \mathcal{L}(\mathcal{H}(g), \mathcal{M}\ominus\mathcal{N})$ such that $T = XS(g)X^{-1}$.

By this last equality and by the relation $WS(\theta)^* = TW$, we infer that $WS(\theta)^* = XS(g)X^{-1}W$, and thus that $(X^{-1}W)S(\theta)^* = S(g)(X^{-1}W)$, so that $S(\theta)^*$ and $S(g)$ are similar. Lastly, observe that by Proposition 3.1.9 $m_{S(\theta)^*} \equiv \tilde{m}_{S(\theta)}$; since, by Proposition 4.1.6, $m_{S(\theta)^*} \equiv m_{S(g)}$, we conclude that $\tilde{\theta} \equiv \tilde{m}_{S(\theta)} \equiv m_{S(g)} \equiv g$, and thus $S(\theta)^*$ is similar to $S(\tilde{\theta})$. □

Before dealing with multiplicity-free operators, we study in more detail the Jordan blocks. First notice that if θ is an inner function, then the vector $P_{\mathcal{H}(\theta)}\mathbf{1}$ is $R(\Omega)$-cyclic for $S(\theta)$. Indeed, $\mathbf{1}$ is $R(\Omega)$-cyclic for S and $S(\theta)P_{\mathcal{H}(\theta)} = P_{\mathcal{H}(\theta)}S$. The operator $S(\theta)$ has many other $R(\Omega)$-cyclic vectors as we shall see shortly.

If $h \in H^2(\Omega)$ and ϕ is an inner function, we say that ϕ divides h ($\phi|h$) if h can be written as $h = \phi g$ for some g in $H^2(\Omega)$. Moreover, by $\phi \wedge h$ we indicate the unique (up to equivalence) inner function which divides ϕ and h and is divisible by any other inner function dividing ϕ and h. Its existence can be proved as in Proposition 2.3.4 (i) using the fully invariant subspace $\phi H^2(\Omega) \oplus hH^\infty(\Omega)$ of $H^2(\Omega)$.

PROPOSITION 4.1.14. *Let θ be a non-trivial inner function.*
(i) For all $h \in \mathcal{H}(\theta)$ we have $m_h \equiv \frac{\theta}{h\wedge\theta}$.
(ii) Every $R(\Omega)$-invariant subspace $\mathcal{M}$ of $S(\theta)$ has the form $\varphi H^2(\Omega) \ominus \theta H^2(\Omega)$ for some inner divisor φ of θ. We have $\varphi H^2(\Omega) \ominus \theta H^2(\Omega) = \ker(\theta/\varphi)(S(\theta)) =$ ran$(\varphi(S(\theta)))$.

(iii) If $\mathcal{M} = \varphi H^2(\Omega) \ominus \theta H^2(\Omega)$ is an $R(\Omega)$-invariant subspace for $S(\theta)$, then $S(\theta)_{|\mathcal{M}}$ is similar to $S(\theta/\varphi)$, and the compression of $S(\theta)$ to $\mathcal{H}(\theta) \ominus \mathcal{M} = \mathcal{H}(\varphi)$ coincides with $S(\varphi)$.
(iv) $h \in \mathcal{H}(\theta)$ is $R(\Omega)$-cyclic for $S(\theta)$ if and only if $\theta \wedge h \equiv \mathbf{1}$.

PROOF. (i) Set $u = m_h$ and $v = \frac{\theta}{h\wedge\theta}$. We have $v(S(\theta))h = P_{\mathcal{H}(\theta)}\theta\frac{h}{h\wedge\theta} = 0$, and consequently $u|v$. Conversely, we know that $u(S(\theta))h = 0$ so that $uh = \theta g$ for some $g \in H^2(\Omega)$. Since $u|\theta$, it follows that $h = (\theta/u)g$ and so $(\theta/u)|h$. Since $(\theta/u)|\theta$ obviously, we have $(\theta/u)|(h \wedge \theta)$, or, equivalently, $v = \frac{\theta}{h\wedge\theta}|u$.

(ii) If $\mathcal{M}$ is an $R(\Omega)$-invariant subspace of $S(\theta)$, then $\mathcal{M} \oplus \theta H^2(\Omega)$ is $R(\Omega)$-invariant for S. So, by Theorem 2.2.14, there exists an inner function φ such that $\mathcal{M}\oplus\theta H^2(\Omega) = \varphi H^2(\Omega)$, that is, $\mathcal{M} = \varphi H^2(\Omega)\ominus\theta H^2(\Omega)$. Clearly $\varphi|\theta$ since $\varphi H^2(\Omega) \supseteq \theta H^2(\Omega)$. If $h \in \varphi H^2(\Omega) \ominus \theta H^2(\Omega)$, obviously $\varphi|h$ so that $\frac{\theta}{h\wedge\theta}|\frac{\theta}{\varphi}$ and $(\theta/\varphi)(S(\theta))h = 0$ by (i). Conversely, if $(\theta/\varphi)(S(\theta))h = 0$, we have $\frac{\theta}{h\wedge\theta}|\frac{\theta}{\varphi}$, which implies that $\varphi|h\wedge\theta$ and hence $\varphi|h$; thus $h \in \varphi H^2(\Omega)\cap\mathcal{H}(\theta) = \varphi H^2(\Omega)\ominus\theta H^2(\Omega)$. For the second equality note that if $\varphi|\theta$, then $\varphi(S(\theta))\mathcal{H}(\theta) = P_{\mathcal{H}(\theta)}\varphi(S)\mathcal{H}(\theta) = P_{\mathcal{H}(\theta)}\varphi(S)H^2(\Omega) = P_{\mathcal{H}(\theta)}\varphi H^2(\Omega) = \varphi H^2(\Omega) \ominus \theta H^2(\Omega)$.

(iii) The space $\mathcal{H}(\varphi) = \mathcal{H}(\theta)\ominus\mathcal{M}$ is $R(\Omega^*)$-invariant for $S(\theta)^*$ and $S(\theta)^*_{|\mathcal{H}(\varphi)} = (S^*_{|\mathcal{H}(\theta)})_{|\mathcal{H}(\varphi)}$. But this last operator is equal to $S^*_{|\mathcal{H}(\varphi)} = S(\varphi)^*$. Thus the compression of $S(\theta)$ to $\mathcal{H}(\varphi)$ is $S(\varphi)$. Let $T = S(\theta)_{|\mathcal{M}}$; T is of class C_0 by Proposition 3.4.3, and $m_T \equiv \theta/\varphi$. Moreover, by Lemma 4.1.12, $T = P_{\mathcal{M}}N_{|\mathcal{M}}$ is similar to $S(g)$ for some inner function g. We conclude by Proposition 4.1.6 that $g \equiv m_T \equiv \theta/\varphi$.

(iv) If $h \in \mathcal{H}(\theta)$ is an $R(\Omega)$-cyclic vector for $S(\theta)$, then $m_h \equiv m_{S(\theta)} \equiv \theta$ so that we have $h \wedge \theta \equiv \mathbf{1}$ by (i). Conversely, if $h \wedge \theta \equiv \mathbf{1}$, (ii) shows that h does not belong to any proper $R(\Omega)$-invariant subspace of $S(\theta)$ and thus h is $R(\Omega)$-cyclic. □

COROLLARY 4.1.15. *Every $R(\Omega)$-invariant subspace of $S(\theta)$ is hyperinvariant.*

COROLLARY 4.1.16. *Let θ be inner. Then, for all $u \in H^\infty(\Omega)$, we have $\ker u(S(\theta)) = (\frac{\theta}{u\wedge\theta})H^2(\Omega)\ominus\theta H^2(\Omega)$, $(\operatorname{ran} u(S(\theta)))^- = (u\wedge\theta)H^2(\Omega)\ominus\theta H^2(\Omega)$, $S(\theta)_{|\ker u(S(\theta))}$ is similar to $S(u\wedge\theta)$ and $S(\theta)_{|(\operatorname{ran} u(S(\theta)))^-}$ is similar to $S(\frac{\theta}{u\wedge\theta})$.*

PROOF. By Theorem 3.4.6, $\ker u(S(\theta)) = \ker(u\wedge\theta)(S(\theta))$, and then by Proposition 4.1.14 (ii) we have that $\ker(u \wedge \theta)(S(\theta)) = \frac{\theta}{u\wedge\theta}H^2(\Omega) \ominus \theta H^2(\Omega)$. We also have by Proposition 3.4.6

$$\mathcal{H}(\theta)\ominus(\operatorname{ran} u(S(\theta)))^- = \ker u(S(\theta))^* = \ker(u\wedge\theta)(S(\theta))^* = \mathcal{H}(\theta)\ominus(\operatorname{ran}(u\wedge\theta)(S(\theta))).$$

Thus Proposition 4.1.14 (ii) implies that $(\operatorname{ran} u(S(\theta)))^- = \operatorname{ran}(u \wedge \theta)(S(\theta)) = (u \wedge \theta)H^2(\Omega) \ominus \theta H^2(\Omega)$. The last two statements of the corollary follow from Proposition 4.1.14 (iii). □

COROLLARY 4.1.17. *Let θ be inner. The set of $R(\Omega)$-cyclic vectors for $S(\theta)$ is a dense G_δ in $\mathcal{H}(\theta)$.*

THEOREM 4.1.18. *Let θ be an inner function and $\mathcal{M} \subseteq \mathcal{H}(\theta)$ be an invariant subspace for $S(\theta)$. Then $\mathcal{M}$ is $R(\Omega)$-invariant for $S(\theta)$.*

PROOF. If $\mathcal{M}$ is invariant for $S(\theta)$, then $\mathcal{N} = \mathcal{M} \oplus \theta H^2(\Omega)$ is invariant for S; indeed, if $x \in \mathcal{M}$, we have $Sx = S(\theta)x + P_{\theta H^2(\Omega)}Sx$. We set $D_j = \{z \in \mathbf{C} : |z - z_j| < r_j\}$ for $j = 1, \dots, m$, and $D_0 = \{z \in \mathbf{C} : |z| > 1\}$.

If $\mathcal{N}$ is not $R(\Omega)$-invariant for S, then there exists a set $A \subseteq \{0, 1, \dots, m\}$ with $0 \in A$ such that $\mathcal{N}$ is invariant under multiplication by any rational function with poles in D_j for every $j \in A$, and $\mathcal{N}$ is not invariant under multiplication by $(z - \zeta)^{-1}$ for every $j \in B = \{0, 1, \dots, m\} - A$ and for every $\zeta \in D_j$ (note that $B \neq \emptyset$). Let $\Omega_B = \mathbf{C} - \cup_{j\in B}\overline{D_j}$ and , for $j \in B$, let $I_j \in H^\infty(\Omega_j)$ be an inner function with modulus equal to 1 a.e. on Γ_j. Following the notation of [3] we use $\mathbf{I} = (I_j)_{j\in B}$ to denote a tuple of such inner functions. By Theorem 6.1 in [3] there exist an inner tuple $\mathbf{I} = (I_j)_{j\in B}$, an outer function $F \in H^2(\Omega_B)$, and a rational function p of the form $p(z) = \prod_{j\in B}(z - z_j)^{m_j}$ for some $m_j \in \mathbf{Z}$, such that

$$\mathcal{N} = \theta' pF\mathcal{N}_{\mathbf{I}},$$

where $\theta' \equiv \wedge\mathcal{N}$ and $\mathcal{N}_{\mathbf{I}} = \{g \in H^2(\Omega) : (g/I_j)_{|\Gamma_j} \in H^2(D_j), \forall j \in B\}$.

If $f \in \mathcal{N}$, then there exists $g \in \mathcal{N}_{\mathbf{I}}$ such that $f = \theta' pFg$. Fix $j \in B$; since $(f/I_j\theta' pF)_{|\Gamma_j} = (g/I_j)_{|\Gamma_j}$, we have that $(f/I_j\theta' pF)_{|\Gamma_j} \in H^2(D_j)$, an thus $(f/I_j\theta' pF)_{|\mathbf{T}}(z_j + r_j z) \in H^2$. Since $\theta H^2(\Omega) \subseteq \mathcal{N}$, then for all $n \in \mathbf{Z}$ we have $f_n = (\frac{z-z_j}{r_j})^n\theta f \in \mathcal{N}$, and thus $\frac{f_n}{I_j\theta' pF}(z_j + r_j z) = \frac{\theta f}{I_j\theta' pF}(z_j + r_j z)z^n = g_n(z) \in H^2$. By Theorem 3.5 in [13] $\int_{\mathbf{T}} g_n(z)dz = 0$, i.e.,

$$\int_{\mathbf{T}} \frac{\theta f}{I_j\theta' pF}(z_j + r_j z)z^n dz = 0.$$

Hence $\frac{\theta f}{I_j\theta' pF}(z_j + r_j z) = 0$ a.e. on $\mathbf{T}$, and thus $\frac{\theta f}{I_j\theta' pF}(z) = 0$ a.e. on Γ_j. Since θ is inner we conclude that $f = 0$ a.e. on Γ_j, and thus we have that f is the zero function in $H^2(\Omega)$. Since $\mathcal{N} \neq 0$ we infer that $\mathcal{N}$ must be $R(\Omega)$-invariant for S, and consequently $\mathcal{M}$ is $R(\Omega)$-invariant for $S(\theta)$. □

4.2. Multiplicity-free operators. We start now studying multiplicity-free operators. The adjoint of a multiplicity-free operator T is not usually multiplicity-free; our first task will be to show that T^* is multiplicity-free if T is of class C_0 and $\mu_T = 1$.

We recall that an operator $S \in \mathcal{L}(H)$ is said to be *subnormal* if there exists a Hilbert space $K \supseteq H$ and a normal operator $N \in \mathcal{L}(K)$ such that $N(H) \subseteq H$ and $N_{|H} = S$. The normal extension N is said to be *minimal* if the space K is the smallest reducing subspace for N containing H.

THEOREM 4.2.1. *Let $T \in \mathcal{L}(H)$ be an operator of class C_0 such that $\mu_T = 1$. Then $S(m_T) \prec T$.*

PROOF. Since similarity is a stronger relation then quasisimilarity, by Proposition 3.2.6 we can without loss of generality assume that T has a minimal normal boundary dilation $M \in \mathcal{L}(K)$, for some Hilbert space K. Let $x \in H$ be an $R(\Omega)$-cyclic vector for T and let K_0 be the reducing subspace for M generated by x; let also $K_+ = \bigvee\{r(M)x : r \in R(\Omega)\}$, $M_+ = M_{|K_+}$ and $M_0 = M_{|K_0}$. Then clearly $K_+ \subseteq K_0$ and M_+ is a subnormal operator having M_0 as minimal normal extension; besides, $\sigma(M_+) \subseteq \overline{\Omega}$, since K_+ is an $R(\Omega)$-invariant subspace for M.

By Proposition 1.1 in [5] M_+ can be written as $M_{+1} \oplus M_{+2}$ where M_{+1} is a normal operator and M_{+2} is a pure operator (i.e., such that there is no non-zero invariant subspaces $\mathcal{R}$ for M_{+2} with the property that the restriction of M_{+2} to $\mathcal{R}$ is normal). M_{+1} is defined on the closed linear span of all subspaces of K_+ reducing for M_0. Let $\mathcal{R} \subseteq K_+$ be a reducing subspace for M_0; then $M_{0|\mathcal{R}} = M_{|\mathcal{R}}$ is normal and clearly $\sigma(M_{|\mathcal{R}}) \subseteq \Gamma$. Set $P = P_{H|\mathcal{R}} : \mathcal{R} \to H$; then $PM_{|\mathcal{R}} = TP$ and thus $Pu(M_{|\mathcal{R}}) = u(T)P$ for all $u \in H^\infty(\Omega)$. Note that by Theorem 3.2.4 the spectral measure of M is absolutely continuous relative to arc-length so that $u(M_{|\mathcal{R}})$ makes sense for all u in $H^\infty(\Omega)$. Then $Pm_T(M_{|\mathcal{R}}) = m_T(T)P = 0$, and, since $m_T(M_{|\mathcal{R}})$ is invertible, we get $P = 0$ and thus $\mathcal{R} \subseteq H^\perp$. Since M is the minimal normal extension of T, we conclude that $\mathcal{R} = 0$ and thus that M_+ is pure. Hence by Proposition 3.3 in [5] the scalar spectral measure of M_0 is mutually absolutely continuous with respect to arc-length on Γ.

Since M_0 has spectral multiplicity one, then there exists a unitary operator $U : L^2(\sigma(M_0), \omega) \to K_0$, such that $M_0 = UNU^*$, where N is the operator of multiplication by z on $L^2(\Gamma, \omega)$. Then $U^*(K_+) = \mathcal{M} \subseteq L^2(\Gamma, \omega)$ is invariant for multiplication of functions in $R(\Omega)$, since K_+ is $R(\Omega)$-invariant for M_0. Suppose $\mathcal{M}$ is reducing for N; then K_+ is reducing for M and $K_0 = K_+$. If $X = P_{H|K_+}$, then $XM_+ = TX$ implies $Xm_T(M_+) = m_T(T)X = 0$, and thus $X = 0$ because $m_T(M_+)$ is invertible. So we have that $K_+ \perp H$, a contradiction. Hence $\mathcal{M}$ is not reducing for N, and by Theorem 1 in [31] we conclude that $\mathcal{M} = uH^2(\Omega)$ for some measurable invertible function u in $L^\infty(\Gamma, \omega)$.

Let us denote by $M_u \in \mathcal{L}(H^2(\Omega), uH^2(\Omega))$ the operator of multiplication by u, and define $Z : H^2(\Omega) \to H$ by $Z = XUM_u$, where as before $X = P_{H|K_+}$. Then $Z \in \mathcal{L}(H^2(\Omega), H)$ since u is bounded below and above. Moreover

$$Z(H^2(\Omega)) = X(UuH^2(\Omega)) = X(U\mathcal{M}) = X(K_+),$$

and since $K_+ = \bigvee\{r(M)x : r \in R(\Omega)\}$ with x $R(\Omega)$-cyclic for T, then Z has dense range. Using the relations $TX = XM_+$ and $M_0U = UN$, we get

$$TZ = TXUM_u = XM_+UM_u = XM_0UM_u$$

$$= XUNM_u = XUM_uN_{|H^2(\Omega)} = XUM_uS = ZS.$$

Hence $TZ = ZS$, and thus $\ker Z$ is a fully invariant subspace of $H^2(\Omega)$. By Theorem 2.2.14 we infer that there exists an inner function θ such that $\ker Z = \theta H^2(\Omega)$. Let $Y = Z_{|\mathcal{H}(\theta)}$; then Y is a quasiaffinity, and since $S(\theta) = P_{\mathcal{H}(\theta)}S_{|\mathcal{H}(\theta)}$ and $Y = Z_{|\mathcal{H}(\theta)}$ with $\ker Z = \theta H^2(\Omega)$, we conclude that $TY = YS(\theta)$, and thus $S(\theta)$ is a quasiaffine transform of T. By Theorem 4.1.6 we reach the conclusion that $\theta \equiv m_T$ and so $S(m_T)$ is a quasiaffine transform of T. □

PROPOSITION 4.2.2. *Let T be an operator of class C_0. If T is multiplicity-free then $S(m_T) \prec T$; if T^* is multiplicity-free then $T \prec S(m_T)$.*

PROOF. The first part of the proposition is exactly Theorem 4.2.1. If T^* is multiplicity-free, from Theorem 3.1.9 we have $S(\tilde{m}_T) = S(m_{T^*}) \prec T^*$, and therefore we also get $T \prec S(\tilde{m}_T)^*$. On the other hand, by Theorem 4.1.13, $S(\tilde{m}_T)^* \prec S((\tilde{m}_T)^\sim) = S(m_T)$ and thus $T \prec S(m_T)$. □

THEOREM 4.2.3. *For every operator T of class C_0, the following are equivalent.*
(i) T is multiplicity-free.
(ii) T^* is multiplicity-free.
(iii) T is quasisimilar to $S(m_T)$.

PROOF. It will suffice to prove that (ii)⇒(i). Indeed, it would then follow that (i)⇒(ii) by simmetry. Further if (i) and (ii) are satisfied, $T \sim S(m_T)$ by the last proposition. Conversely, if we suppose that $T \prec S(m_T)$, then (i) follows by Lemma 4.1.9.

Assume therefore that $T \in \mathcal{L}(H)$ and T^* is multiplicity-free. By Proposition 4.2.2 we can choose a quasiaffinity $X \in \mathcal{L}(H, \mathcal{H}(m_T))$ such that

$$S(m_T)X = XT. \tag{1}$$

Let $x \in H$ be a T-maximal vector, i.e. $m_x \equiv m_T$. If $\mathcal{M} = \bigvee\{r(T)x : r \in R(\Omega)\}$, then $\mathcal{M}$ is $R(\Omega)$-invariant for T and $m_{T_{|\mathcal{M}}} \equiv m_T$. Thus $T_{|\mathcal{M}}$ is a multiplicity-free and a second application of the last proposition yields a one-to-one operator $Y : \mathcal{H}(m_T) \to H$ such that $Y(\mathcal{H}(m_T))$ is dense in $\mathcal{M}$ and

$$TY = YS(m_T). \tag{2}$$

Relations (1) and (2) show that $XY \in \{S(m_T)\}'$ and clearly XY is one-to-one. So by Corollary 4.1.3, there exists $u \in H^\infty(\Omega)$ such that

$$XY = u(S(m_T)), \tag{3}$$

and by Proposition 3.4.6 $u \wedge m_T \equiv \mathbf{1}$. A calculation using (1) and (3) shows that $X(YX - u(T)) = 0$, and since X is one-to-one we have $YX = u(T)$. Moreover, by Proposition 3.4.6, $u(T)$ is a quasiaffinity, since $u \wedge m_T \equiv \mathbf{1}$. Hence $H = (u(T)H)^- \subseteq (Y\mathcal{H}(m_T))^- \subseteq \mathcal{M}$, so that $\mathcal{M} = H$. □

The above argument actually shows more. The following result is an immediate consequence of it.

COROLLARY 4.2.4. *Let T be multiplicity-free of class C_0. Then a vector $x \in H$ is $R(\Omega)$-cyclic for T if and only if x is T-maximal. In particular the set of $R(\Omega)$-cyclic vectors for T is a dense G_δ in H.*

COROLLARY 4.2.5. *Every restriction of a multiplicity-free operator of class C_0 to an $R(\Omega)$-invariant subspace is multiplicity-free.*

PROOF. Let T be multiplicity-free and $\mathcal{M}$ be an $R(\Omega)$-invariant subspace for T. If x is an $R(\Omega^*)$-cyclic vector for T^*, then $P_{\mathcal{M}}x$ is $R(\Omega^*)$-cyclic for $(T_{|\mathcal{M}})^*$. Thus $(T_{|\mathcal{M}})^*$, and hence $T_{|\mathcal{M}}$, is multiplicity-free. □

Many of the properties of Jordan blocks are inherited in some form by multiplicity-free operators. We recall that for an operator T satisfying hypothesis (h), the closed operator $(u/v)(T) = v(T)^{-1}u(T)$ is defined whenever $u \in H^\infty(\Omega)$ and $v \in K_T^\infty(\Omega)$. We denote by $\mathcal{F}_T$ the set of bounded operators of the form $(u/v)(T)$.

LEMMA 4.2.6. *Let $T \in \mathcal{L}(H)$ and $T' \in \mathcal{L}(H')$ be two multiplicity-free C_0 operators with $m_T \equiv m_{T'}$. Then an operator $A \in \mathcal{L}(H, H')$ such that $AT = T'A$ is one-to-one if and only if it has dense range.*

PROOF. If $\theta \equiv m_T$, it follows from Theorem 4.2.3 that we can find quasiaffinities X, Y such that $XT' = S(\theta)X$ and $TY = YS(\theta)$. Then the product XAY commutes with $S(\theta)$, and hence Corollary 4.1.3 yields a function $u \in H^\infty(\Omega)$ such that $XAY = u(S(\theta))$. If A is either one-to-one or has dense range, then $u(S(\theta))$ has the same property, and therefore, by Proposition 3.4.6, $u \wedge \theta \equiv \mathbf{1}$. The relations $AYX = u(T')$ and $YXA = u(T)$, imply that $AYX = u(T')$ and $YXA = u(T)$, since Y and X are quasiaffinities. We infer from these last relations that $\operatorname{ran} A \supseteq \operatorname{ran} u(T')$ and $\ker A \subseteq \ker u(T)$. The condition $u \wedge \theta \equiv \mathbf{1}$ implies that $u(T)$ and $u(T')$ are quasiaffinities by Proposition 3.4.6, and thus A is a quasiaffinity. □

PROPOSITION 4.2.7. *If T is a multiplicity-free operator of class C_0, then $\mathcal{F}_T = \{T\}'$.*

PROOF. Let θ, X, Y be as in the proof of Lemma 4.2.6 with $T' = T$, and let $A \in \{T\}'$. If $A = I$, the proof implies the existence of $v \in H^\infty(\Omega)$ such that $v \wedge \theta \equiv \mathbf{1}$ and $YX = v(T)$; if A is arbitrary, we deduce the existence of $u \in H^\infty(\Omega)$ such that $YXA = u(T)$, or $v(T)A = u(T)$. It follows that $A = (u/v)(T)$. □

Something more than the equality $\{T\}' = \mathcal{F}_T$ is true; indeed, there exists $v \in H^\infty(\Omega)$ such that $v \wedge m_T \equiv \mathbf{1}$ and every $A \in \{T\}'$ can be written as $A = (u/v)(T)$ for some $u \in H^\infty(\Omega)$.

The rationally invariant subspaces of multiplicity-free operators of class C_0 have a classification, analogous to that of rationally invariant subspaces for Jordan blocks.

THEOREM 4.2.8. *Let T be an operator of class C_0. The following are equivalent.*
(i) T is multiplicity-free.
(ii) For every inner divisor θ of m_T, there exists a unique $R(\Omega)$-invariant subspace $\mathcal{M}$ for T such that $m_{T|\mathcal{M}} \equiv \theta$.
(iii) There do not exist distinct $R(\Omega)$-invariant subspaces $\mathcal{M}$ and $\mathcal{M}'$ for T such that $T_{|\mathcal{M}} \prec T_{|\mathcal{M}'}$.
(iv) There do not exist proper $R(\Omega)$-invariant subspaces $\mathcal{M}$ for T such that $m_{T|\mathcal{M}} \equiv m_T$.
If T is multiplicity-free, then the unique $R(\Omega)$-invariant subspace in (ii) is given by $\mathcal{M} = \ker\theta(T) = (\operatorname{ran}(m_T/\theta)(T))^-$.

PROOF. (i)⇒(ii) Assume that T is multiplicity-free, $\mathcal{M}$ is an $R(\Omega)$-invariant subspace for T and $\theta \equiv m_{T|\mathcal{M}}$. The operators $T' = T_{|\mathcal{M}}$ and $T'' = T_{|\ker\theta(T)}$ are multiplicity-free by Corollary 4.2.5 and they satisfy the relation $T''J = JT'$, where $J : \mathcal{M} \to \ker\theta(T)$ is the inclusion operator. By Lemma 4.2.6, J must have dense range so that $\mathcal{M} = J(\mathcal{M}) = \ker\theta(T)$.

(ii)⇒(iv) Obvious.

(iv)⇒(i) Let x be a T-maximal vector and let $\mathcal{M} = \bigvee\{r(T)x : r \in R(\Omega)\}$; then $m_{T_{|\mathcal{M}}} \equiv m_x \equiv m_T$. Therefore $\mathcal{M} = H$ by (iv).

(iii)⇒(i) If x and x' are T-maximal vectors, $\mathcal{M} = \bigvee\{r(T)x : r \in R(\Omega)\}$ and $\mathcal{M}' = \bigvee\{r(T)x' : r \in R(\Omega)\}$, then $T_{|\mathcal{M}}$ and $T_{|\mathcal{M}'}$ are multiplicity-free operators having the same minimal function. By Theorem 4.2.3 we deduce, using the transitivity of quasisimilarity, that $T_{|\mathcal{M}} \prec T_{|\mathcal{M}'}$. If (iii) holds we must then have $\mathcal{M}' = \mathcal{M}$ and in particular $x' \in \mathcal{M}$. We conclude that $\mathcal{M}$ contains every T-maximal vector and hence $\mathcal{M} = H$, since the set of T-maximal vectors is dense. Thus x is $R(\Omega)$-cyclic.

(ii) ⇒(iii) This is obvious because $T_{|\mathcal{M}} \prec T_{|\mathcal{M}'}$ implies that $m_{T_{|\mathcal{M}}} \equiv m_{T_{|\mathcal{M}'}}$.

The last assertion of the theorem follows because both the operators $T_{|\ker\theta(T)}$ and $T_{|(\operatorname{ran}(m_T/\theta)(T))^-}$ have minimal function θ. □

COROLLARY 4.2.9. *Every $R(\Omega)$-invariant subspace of a multiplicity-free operator of class C_0 is hyperinvariant.*

4.3. The classification theorem. The classification of C_0 operators over finitely connected regions with arbitrary multiplicity is similar to that of C_0 operators over the disk (see [7] chapter 3). The main difference is that in this new context we need to use $R(\Omega)$-invariant subspaces rather than simply invariant subspaces, since we need them to be invariant for the functional calculus representation.

The splitting theorem below already suggests how the classification of general C_0 operators will look.

THEOREM 4.3.1. **The splitting theorem** *Let $T \in \mathcal{L}(H)$ be an operator of class C_0, $x \in H$ a T-maximal vector and $\mathcal{K} = \bigvee\{r(T)x : r \in R(\Omega)\}$. Then there exists an $R(\Omega)$-invariant subspace $\mathcal{M}$ for T such that $\mathcal{K} \vee \mathcal{M} = H$ and $\mathcal{K} \cap \mathcal{M} = \{0\}$.*

PROOF. The operator $T_1 = T_{|\mathcal{K}}$ is multiplicity-free, and by virtue of Theorem 4.2.3 there exists a vector $k \in \mathcal{K}$ which is $R(\Omega^*)$-cyclic for T_1^*. Let us denote $\mathcal{K}' = \bigvee\{r(T^*)k : r \in R(\Omega^*)\}$, where $\mathcal{M} = H \ominus \mathcal{K}'$, and define $T_2 \in \mathcal{L}(\mathcal{K}')$ by $T_2^* = T^*_{|\mathcal{K}'}$. Since $\mathcal{K}'$ is $R(\Omega^*)$-invariant for T^*, we have $T_2 P_{\mathcal{K}'} = P_{\mathcal{K}'} T$, and hence the operator $X \in \mathcal{L}(\mathcal{K}, \mathcal{K}')$ defined by $X = P_{\mathcal{K}'|\mathcal{K}}$, satisfies the relation $T_2 X = X T_1$.

Since X^* has dense range, then $(m_{T_2}(T_1))^* X^* = X^*(m_{T_2}(T_2))^* = 0$ implies $m_{T_2}(T_1) = 0$. So $m_T \equiv m_{T_1} | m_{T_2}$, and since by Proposition 3.4.2 $m_{T_2} | m_T$, we deduce that $m_{T_1} \equiv m_{T_2}$. Lemma 4.2.6 now implies that X must be a quasiaffinity, and the theorem follows from the relations $\mathcal{K} \cap \mathcal{M} = \ker X = \{0\}$, and $H \ominus (\mathcal{K} \vee \mathcal{M}) = \ker X^* = \{0\}$. □

By means of this theorem it is possible to prove the converse to Proposition 4.2.7.

THEOREM 4.3.2. *For every operator T of class C_0 the following are equivalent.*
(i) T is multiplicity-free.
(ii) $\{T\}'$ is commutative.
(iii) $\{T\}' = \mathcal{F}_T$.

PROOF. We already know by Proposition 4.2.7 that (i) implies (iii), and (iii) trivially implies (ii). It remains to show that the commutant $\{T\}'$ of a non multiplicity-free operator T of class C_0 cannot be commutative.

Let us define, as in the proof of Theorem 4.3.1, $\mathcal{K} = \bigvee\{r(T)x : r \in R(\Omega)\}$ and $\mathcal{M} = H \ominus \mathcal{K}'$, where $\mathcal{K}' = \bigvee\{r(T^*)x : r \in R(\Omega^*)\}$, and x is T-maximal. The operator $X = P_{\mathcal{K}'|\mathcal{K}}$ is a quasiaffinity and if $T_1 = T_{|\mathcal{K}}$ and $T_2^* = T^*_{|\mathcal{K}'}$, then $XT_1 = T_2X$, T_1 and T_2 are multiplicity-free operators of class C_0 and $m_{T_1} \equiv m_{T_2} \equiv m_T$. Thus T_1 and T_2 are quasisimilar, since they are both quasisimilar to $S(m_T)$ by Theorem 4.2.3. Let Y be a quasiaffinity such that $YT_2 = T_1Y$, and define $A \in \{T\}'$ by $A = YP_{\mathcal{K}'}$. We clearly have $\ker A = \ker P_{\mathcal{K}'} = \mathcal{M}$ and $(AH)^- = \mathcal{K}$.

Suppose we can find a nonzero operator Z with the property that $ZT_2 = T_{|\mathcal{M}}Z$. Then the operator $B \in \{T\}'$ defined by $B = ZP_{\mathcal{K}'}$, is such that $AB \neq BA$. Thus, in order to show that $\{T\}'$ is not commutative, it suffices to show the existence of such a Z.

Since $\mathcal{M}$ is $R(\Omega)$-invariant for T and is not the zero subspace, then $T_{|\mathcal{M}}$ has a nonzero $R(\Omega)$-cyclic subspace $\mathcal{M}_1$, and it would suffice to find $Z \neq 0$ such that $ZT_2 = T_{|\mathcal{M}_1}Z$. Finally, if $\theta \equiv m_T \equiv m_{T_2}$ and $\theta' \equiv m_{T_{|\mathcal{M}_1}}$, then $T_{|\mathcal{M}_1}$ is C_0 and multiplicity-free, so that $T_{|\mathcal{M}_1} \sim S(\theta')$, $T_2 \sim S(\theta)$ and $\theta'|\theta$. Therefore we can as well prove that there are non-zero operators V such that $VS(\theta) = S(\theta')V$ whenever $\theta'|\theta$ and θ' is non-trivial. The existence of such operators follows from the observations before Theorem 4.1.2. □

If $T \in \mathcal{L}(H)$ is an operator and γ a cardinal number, we denote by $T^{(\gamma)}$ the direct sum of γ copies of T acting on the direct sum $H^{(\gamma)}$ of γ copies of H.

It follows from the splitting theorem that if T is an operator of class C_0, then there exists another operator T' of class C_0 such that $S(m_T) \oplus T' \prec T$. Indeed, let $\mathcal{K}$ and $\mathcal{M}$ be as in Theorem 4.3.1. Then we have $T_{|\mathcal{K}} \oplus T_{|\mathcal{M}} \prec T$, where the quasiaffinity $X : \mathcal{K} \oplus \mathcal{M} \to H$ is defined by $X(u \oplus v) = u + v$. Moreover, X is one-to-one since $\mathcal{K} \cap \mathcal{M} = \{0\}$, and has dense range since $\mathcal{K} \vee \mathcal{M} = H$. The assertion follows because $T_{|\mathcal{K}}$ is multiplicity-free and $m_{T_{|\mathcal{K}}} \equiv m_T$, so that $T_{|\mathcal{K}}$ is quasisimilar to $S(m_T)$ by Theorem 4.2.3. Hence for a given operator T of class C_0 we can find inner functions $\theta_0, \theta_1, \ldots, \theta_n$ and another operator T' of class C_0 such that $\theta_{j+1}|\theta_j$, $0 \leq j \leq n-1$, and $S(\theta_0) \oplus S(\theta_1) \oplus \cdots \oplus S(\theta_n) \oplus T' \prec T$. An inductive application of this observation (transfinite if the space is not separable) yields a direct sum of Jordan blocks that is actually quasisimilar to T. For reasons of uniqueness we allow only certain kinds of direct sums of Jordan blocks that we call Jordan operators.

An important step to the classification theorem is the proof that for any inner function θ and for any natural number n we have $\mu_{S(\theta)^{(n)}} = n$. Even if the proof of the following lemma goes like in the case of the disk, the proof of Theorem 4.3.4 is different, and it uses the theory of bundle shifts developed in [5].

LEMMA 4.3.3. *Let θ be a non-trivial inner function and n, k be two natural numbers.*
(i) If there exists a one-to-one operator X such that $XS(\theta)^{(k)} = S(\theta)^{(n)}X$, then $k \leq n$.

(ii) If there exists an operator X with dense range and such that $S(\theta)^{(n)}X = XS(\theta)^{(k)}$, then $k \geq n$.

PROOF. (i) Let $P_i : \mathcal{H}(\theta)^{(n)} \to \mathcal{H}(\theta)$ and $Q_j : \mathcal{H}(\theta)^{(k)} \to \mathcal{H}(\theta)$, $1 \leq i \leq n$, $1 \leq j \leq k$, be the natural projections, and assume that X satisfies $XS(\theta)^{(k)} = S(\theta)^{(n)}X$. The operators $P_iXQ_j^*$ commute with $S(\theta)$, and by virtue of Corollary 4.1.3 we can find $a_{ij} \in H^\infty(\Omega)$ such that $P_iXQ_j^* = a_{ij}(S(\theta))$, $1 \leq i \leq n$, $1 \leq j \leq k$. Thus the operator X can be written as

$$X\big(\oplus_{j=1}^k h_j\big) = \bigoplus_{i=1}^n P_{\mathcal{H}(\theta)}\Big(\sum_{j=1}^k a_{ij}h_j\Big),$$

$h_j \in \mathcal{H}(\theta)$, $1 \leq j \leq k$. Assume now that X is one-to-one; in this case not all the functions a_{ij} can be divisible by θ. Therefore there exists a minor of maximum rank of the matrix $[a_{ij}]_{i,j}$ that is not divisible by θ, and there is no loss of generality in assuming that the minor is $|a_{ij}|_{1\leq i,j\leq r}$, where $r \leq \min(k,n)$. Assume that $k > n$ and write the determinant

$$\begin{vmatrix} a_{11} & a_{12} & \cdots & a_{1,r+1} \\ a_{21} & a_{22} & \cdots & a_{2,r+1} \\ \vdots & \vdots & \ddots & \vdots \\ a_{r1} & a_{r2} & \cdots & a_{r,r+1} \\ x_1 & x_2 & \cdots & x_{r+1} \end{vmatrix} = \sum_{j=1}^{r+1} x_j u_j,$$

that we developed following the last row. The sum $\sum_{j=1}^{r+1} a_{ij}u_j$ is zero if $1 \leq i \leq r$ and equals a minor of order $r+1$ (hence divisible by θ) if $i > r$. Then the vector $h = \oplus_{j=1}^k h_j \in \mathcal{H}(\theta)^{(k)}$ defined by $h_j = P_{\mathcal{H}(\theta)}u_j$ if $1 \leq j \leq r+1$ and $h_j = 0$ otherwise, satisfies $Xh = 0$ so that $h = 0$. So we have $P_{\mathcal{H}(\theta)}u_{r+1} = 0$, i.e., $u_{r+1} \in \theta H^2(\Omega)$. But $u_{r+1} = |a_{ij}|_{1\leq i,j\leq r}$ was chosen not divisible by θ, a contradiction.

(ii) The proof easily follows from (i) applied to the adjoint X^*, since the operator $S(\theta)^*$ is similar to the Jordan block $S(\tilde{\theta})$ by Theorem 4.1.13. $\square$

THEOREM 4.3.4. *Let θ be a non-trivial inner function and n be a natural number. Then the multiplicity of the operator $S(\theta)^{(n)}$ equals n.*

PROOF. Set $k = \mu_{S(\theta)^{(n)}}$. Since $S(\theta)$ is multiplicity-free, the inequality $k \leq n$ is clear. So it is enough to prove that $k \geq n$. Choose $\mathcal{M} \subseteq \mathcal{H}(\theta)^{(n)}$ such that $\mathrm{card}(\mathcal{M}) = k$ and $\mathcal{H}(\theta)^{(n)} = \bigvee\{r(S(\theta)^{(n)})m : r \in R(\Omega), m \in \mathcal{M}\}$. Let $N^{(n)}$ be the operator defined on $L^2(\Gamma,\omega)^{(n)}$ as direct sum of n-times the operator N of multiplication by z. Let $H_1 = \bigvee\{r(N^{(n)})m : r \in R(\Omega), m \in \mathcal{M}\}$ and H_0 be the reducing subspace for $N^{(n)}$ generated by $\mathcal{M}$; then $H_1 \subseteq H_0 \subseteq L^2(\Gamma,\omega)^{(n)}$. Let $N_1 = N^{(n)}_{|H_1}$, $N_0 = N^{(n)}_{|H_0}$ and $X = P_{\mathcal{H}(\theta)^{(n)}|H_1} : H_1 \to \mathcal{H}(\theta)^{(n)}$. Then X has dense range; indeed

$$\begin{aligned}(X(H_1))^- &= (X(\bigvee\{r(N^{(n)})m : r \in R(\Omega), m \in \mathcal{M}\}))^- \\ &= \bigvee\{Xr(N^{(n)})m : r \in R(\Omega), m \in \mathcal{M}\}\end{aligned}$$

$$= \bigvee\{r(S(\theta)^{(n)})m : r \in R(\Omega), m \in \mathcal{M}\} = \mathcal{H}(\theta)^{(n)}.$$

Moreover, if $x = r(N^{(n)})m$ for some $r \in R(\Omega)$ and $m \in \mathcal{M}$, then

$$XN_1x = XN_1r(N^{(n)})m = Xr(N_1^{(n)})N_1m = r(S(\theta)^{(n)})S(\theta)^{(n)}m,$$

and

$$S(\theta)^{(n)}Xx = S(\theta)^{(n)}Xr(N^{(n)})m = S(\theta)^{(n)}r(S(\theta)^{(n)})m = r(S(\theta)^{(n)})S(\theta)^{(n)}m;$$

hence $XN_1 = S(\theta)^{(n)}X$.

N_1 is a subnormal operator and N_0 is its minimal normal extension; besides $\sigma(N_0) \subseteq \Gamma$, and, N_1 being defined on a subspace of H_0 which is $R(\Omega)$-invariant, we have $\sigma(N_1) \subseteq \overline{\Omega}$. We clearly have $\mu_{N_1} \leq \text{card}(\mathcal{M}) = k$, and, by Lemma 4.1.9, $k \leq \mu_{N_1}$; hence we conclude that $\mu_{N_1} = k$. N_1 is a pure subnormal operator (for the definition of pure subnormal operator cf. the proof of Theorem 4.2.1); indeed, let $\mathcal{R} \subseteq H_1$ be a reducing subspace for N_0, then $N_{1|\mathcal{R}} = N_{0|\mathcal{R}}$ is normal. Let $Y = P_{\mathcal{H}(\theta)^{(n)}|\mathcal{R}} : \mathcal{R} \to \mathcal{H}(\theta)^{(n)}$; then the relation $YN_{0|\mathcal{R}} = S(\theta)^{(n)}Y$ implies

$$Y\theta(N_{0|\mathcal{R}}) = \theta(S(\theta)^{(n)})Y = 0,$$

(note that by Theorem 3.2.4 the spectral measure of N is absolutely continuous relative to arc-length, hence $u(N_{0|\mathcal{R}})$ is defined for all $u \in H^\infty(\Omega)$). Since $\theta(N_{0|\mathcal{R}})$ is invertible, we get that Y is the zero operator and thus that $\mathcal{R} \subseteq (\mathcal{H}(\theta)^{(n)})^\perp$, which is a contradiction since N_0 is the minimal normal extension of N_1. By Theorem 11 in [5], N_1 is unitarily equivalent to a bundle shift T_E over Ω. Since $\mu_{N_1} = k$, then $\mu_{T_E} = k$, and hence, by Theorem 3 in [5], the dimension of the vector bundle E is $j \leq k$. Let now $F = \Omega \times \mathbf{C}^{(j)}$, and let us consider the bundle shift over Ω defined by $T_F = N^{(j)}_{|H^2(\Omega)^{(j)}}$. Since $\dim(F) = \dim(E)$, then by the corollary to Theorem 3 in [5], T_F is similar to T_E, and thus there exists an invertible operator Z such that $ZT_F = N_1Z$.

Let $W = XZ$, then W has dense range and

$$WT_F = XZT_F = XN_1Z = S(\theta)^{(n)}XZ = S(\theta)^{(n)}W;$$

hence $WT_F = S(\theta)^{(n)}W$. So $W\theta(T_F) = \theta(S(\theta)^{(n)})W = 0$, and in conclusion

$$W_{|\theta(T_F)H^2(\Omega)^{(j)}} = W_{|\theta H^2(\Omega)^{(j)}} = 0.$$

Let $K = W_{|\mathcal{H}(\theta)^{(j)}}$; we have $S(\theta)^{(n)}K = WT_{F|\mathcal{H}(\theta)^{(j)}} = WN^{(j)}_{|\mathcal{H}(\theta)^{(k)}} = KS(\theta)^{(j)}$ and K has dense range. By Lemma 4.3.3 $j \geq n$, and thus $k \geq n$. $\square$

We recall that an *ordinal number* is a set α that is transitive with respect to "$\in$", i.e., $\beta \in \gamma$ and $\gamma \in \alpha$ implies $\beta \in \alpha$, and is totally ordered by "$\in$". Thus $0 = \emptyset$, $1 = \{\emptyset\}$, $2 = \{\emptyset, \{\emptyset\}\}$, $\ldots$ and, in general, $\alpha = \{\beta : \beta \in \alpha\}$. We write $\alpha < \beta$ if α and β are ordinals and $\alpha \in \beta$. An ordinal α is a *cardinal number* if it is not equipotent with any smaller ordinal. Thus $0, 1, 2, \ldots$ are cardinal numbers and ω, the first infinite ordinal, is also a cardinal number and is sometimes denoted by $\aleph_0$. Ordinals are well ordered and so the definition

$$\text{card}(\alpha) = \text{ first } \beta \text{ such that } \beta \leq \alpha \text{ and } \beta \text{ is equipotent to } \alpha$$

makes sense; it associates a cardinal number with every ordinal. The axiom of choice implies that any set $\mathcal{M}$ is equipotent to some cardinal α; in this case $\alpha = \operatorname{card}(\mathcal{M})$.

DEFINITION 4.3.5. Let γ be a cardinal number and $\Theta = \{\theta_\alpha : \alpha < \gamma\}$ be a family of inner functions. Then Θ is called a *model function* if $\theta_\alpha | \theta_\beta$ whenever $\operatorname{card}(\beta) \leq \operatorname{card}(\alpha) < \gamma$. The *Jordan operator* $S(\Theta)$ determined by the model function Θ is the C_0 operator defined as

$$S(\Theta) = \bigoplus_{\alpha < \gamma'} S(\theta_\alpha), \quad \gamma' = \min\{\beta : \theta_\beta \equiv \mathbf{1}\}.$$

We will denote by $\mathcal{H}(\Theta)$ the space of $S(\Theta)$.

The condition that Θ be a model function implies that $m_{S(\Theta)} \equiv \theta_0$. The fact that some θ_α might be trivial suggests the following definition.

DEFINITION 4.3.6. Let $\Theta = \{\theta_\alpha : \alpha < \gamma\}$ and $\Theta' = \{\theta'_\alpha : \alpha < \gamma'\}$ be two model functions. We say that Θ and Θ' are *equivalent* (and we write $\Theta \equiv \Theta'$) if $\theta_\alpha \equiv \theta'_\alpha$ for $\alpha < \min\{\gamma, \gamma'\}$ and $\theta_\alpha \equiv \mathbf{1}$ [resp. $\theta'_\alpha \equiv \mathbf{1}$] if $\gamma' \leq \alpha < \gamma$ [resp. $\gamma \leq \alpha < \gamma'$].

It is clear that the relation $\Theta \equiv \Theta'$ implies $S(\Theta) = S(\Theta')$ and thus the number γ is not relevant in the definition of Jordan operators. It may be easier to think that θ_α is defined for every ordinal α, θ_α is trivial if α is large enough, and $S(\Theta) = \oplus_{\alpha<\gamma} S(\theta_\alpha)$, where γ is the first ordinal such that $\theta_\gamma \equiv \mathbf{1}$.

The main goal now is to prove that each quasisimilarity class in $\mathcal{L}(H)$ contains, up to unitarily equivalence, at most one Jordan operator. Then we will show that the quasisimilarity class of an operator of class C_0 contains a Jordan operator. We start determining the functions μ_T and ν_T corresponding with a Jordan operator T.

THEOREM 4.3.7. *Let $\Theta = \{\theta_\alpha : \alpha < \gamma\}$ be a model function. Then*
(i) $\mu_{S(\Theta)} = \min\{\alpha : \theta_\alpha \equiv \mathbf{1}\}$;
(ii) $\nu_{S(\Theta)}(\theta) = \min\{\alpha : \theta_\alpha | \theta\}$ for every inner function θ.

PROOF. Note first that the two minima in the statement are actually cardinal numbers by the definition of model function. The more general statement (ii) is implied by (i). Indeed, for all α, $S(\theta_\alpha)|(\operatorname{ran}\theta(S(\theta_\alpha)))^-$ is similar to $S(\frac{\theta_\alpha}{\theta \wedge \theta_\alpha})$ by Corollary 4.1.16, so that $S(\Theta)|(\operatorname{ran}\theta(S(\Theta)))^-$ is similar to $\bigoplus_{\alpha<\gamma} S(\frac{\theta_\alpha}{\theta \wedge \theta_\alpha})$. Now (ii) follows from (i) if we note that $\{\frac{\theta_\alpha}{\theta \wedge \theta_\alpha} : \alpha < \gamma\}$ is also a model function, and $\theta_\alpha | \theta$ if and only if $\frac{\theta_\alpha}{\theta \wedge \theta_\alpha} \equiv \mathbf{1}$. To prove (i) let $\beta = \min\{\alpha : \theta_\alpha \equiv \mathbf{1}\}$. The direct sum $S(\Theta) = \bigoplus_{\alpha<\beta} S(\theta_\alpha)$ contains exactly β multiplicity-free summands and the inequality $\mu_{S(\Theta)} \leq \beta$ follows at once. If $\beta > \aleph_0$ we have $\beta \leq \dim(\mathcal{H}(\Theta)) \leq \aleph_0 \beta = \beta$, and hence $\dim(\mathcal{H}(\Theta)) = \beta$. The equality $\mu_{S(\Theta)} = \beta$ follows now from the inequalities $\mu_{S(\Theta)} \leq \beta \leq \aleph_0 \mu_{S(\Theta)}$. We still have to treat the case in which $\beta \leq \omega$. First consider the subcase $\beta < \omega$. Then $\theta_{\beta-1}$ is a non-trivial divisor of $\theta_0, \theta_1, \dots, \theta_{\beta-2}$ and therefore, by Proposition 4.1.14 (iii), $S(\theta_{\beta-1})^* = S(\theta_j)^*_{|\mathcal{H}(\theta_{\beta-1})}$, $j = 0, 1, \dots, \beta - 2$. Hence we deduce

$$(S(\theta_{\beta-1})^{(\beta)})^* = S(\Theta)^*_{|(\oplus_{j=0}^{\beta-1}\mathcal{H}(\theta_{\beta-1}))}.$$

Using Theorem 4.3.4, we see that this relation implies $\beta = \mu_{S(\theta_{\beta-1})^{(\beta)}} \le \mu_{S(\Theta)}$ and (i) is proved if $\beta < \omega$. Finally, if $\beta = \omega$, by the preceding case we clearly have $\mu_{S(\Theta)} \ge \mu_{\bigoplus_{j=0}^{n-1} S(\theta_j)} = n$ for every $n < \omega$. Indeed all functions θ_j, $j < \omega$, are non-trivial by thc definition of β. We conclude that $\mu_{S(\Theta)} \ge \omega = \beta$. □

We have shown how to compute $\mu_{S(\Theta)}$ given the model function Θ. It is an easy task to reverse this process.

COROLLARY 4.3.8. *If $\Theta = \{\theta_\alpha : \alpha < \gamma\}$ is a model function then for each ordinal α*

$$\theta_\alpha \equiv \bigwedge\{\theta : \nu_{S(\Theta)}(\theta) \le \text{ card } (\alpha)\}.$$

PROOF. We have $\nu_{S(\Theta)}(\theta_\alpha) = \min\{\beta : \theta_\beta|\theta_\alpha\} \le \alpha$, so that $\nu_{S(\Theta)}(\theta_\alpha) \le$ card(α) (note that $\nu_{S(\Theta)}(\theta_\alpha)$ is a cardinal number), and we conclude that the function $\varphi \equiv \bigwedge\{\theta : \nu_{S(\Theta)}(\theta) \le \text{ card}(\alpha)\}$ divides θ_α. Now let θ be an inner function such that $\nu_{S(\Theta)}(\theta) \le$ card$(\alpha) \le \alpha$. Since $\nu_{S(\Theta)}(\theta) = \min\{\beta : \theta_\beta|\theta\}$, we deduce that $\theta_\alpha|\theta$. Because θ was arbitrary such that $\nu_{S(\Theta)}(\theta) \le$ card(α), we must have $\theta_\alpha|\varphi$. □

THEOREM 4.3.9. **Uniqueness Property** *If Θ and Θ' are two model functions and $S(\Theta) \prec S(\Theta')$, then $\Theta \equiv \Theta'$ and hence $S(\Theta) = S(\Theta')$.*

PROOF. Assume that $\Theta = \{\theta_\alpha\}$ and $\Theta' = \{\theta'_\alpha\}$. If $S(\Theta) \prec S(\Theta')$, Proposition 4.1.11 implies $\nu_{S(\Theta')}(\theta) \le \nu_{S(\Theta)}(\theta)$ for every inner function θ. Thus, for every ordinal α,

$$\{\theta : \nu_{S(\Theta)}(\theta) \le \text{ card}(\alpha)\} \subseteq \{\theta : \nu_{S(\Theta')}(\theta) \le \text{ card}(\alpha)\},$$

so that $\theta'_\alpha \equiv \bigwedge\{\theta : \nu_{S(\Theta')}(\theta) \le \text{ card}(\alpha)\}$ divides $\theta_\alpha \equiv \bigwedge\{\theta : \nu_{S(\Theta)}(\theta) \le \text{ card}(\alpha)\}$. Finally, we have $S(\Theta')^* \prec S(\Theta)^*$ and these two operators are also similar to Jordan operators, determined by the model functions $\{\tilde{\theta}'_\alpha\}$ and $\{\tilde{\theta}_\alpha\}$, respectively. By the first part of the proof we have $\tilde{\theta}_\alpha|\tilde{\theta}'_\alpha$ so that $\theta_\alpha|\theta'_\alpha$. We conclude that $\theta_\alpha \equiv \theta'_\alpha$ and the theorem follows. □

We are about to show that operators of class C_0 can be completely classified using quasisimilarity. An iterated application of the splitting procedure and a density argument yields a proof of the classification theorem for operators acting on a separable space. Separably acting Jordan operators are of the form $\oplus_{j=0}^{\infty} S(\theta_j)$, where $\{\theta_j : j \ge 0\}$ is a sequence of inner functions such that $\theta_{j+1}|\theta_j$ for $j \ge 0$.

THEOREM 4.3.10. *Let T be an operator of class C_0 acting on a separable space H. There exists a Jordan operator $S(\Theta)$ that is quasisimilar to T. Moreover, $S(\Theta)$ is uniquely determined by either of the relations $S(\Theta) \prec T$ and $T \prec S(\Theta)$.*

PROOF. Let $\{h_n\}_{n=0}^{\infty}$ be a dense sequence in H and let $\{k_n\}_{n=0}^{\infty}$ be a sequence in which each h_n is repeated infinitely often. We construct inductively vectors $x_0, x_1, \dots$ in H and $R(\Omega)$-invariant subspaces $\mathcal{M}_{-1}, \mathcal{M}_0, \mathcal{M}_1, \dots$ for T, such that $\mathcal{M}_{-1} = H$ and for $j = 0, 1, \dots$ we have:

$$x_j \in \mathcal{M}_{j-1}, \quad m_{x_j} = m_{T|\mathcal{M}_{j-1}}, \tag{1}$$

$$\mathcal{K}_j \vee \mathcal{M}_j = \mathcal{M}_{j-1}, \quad \mathcal{K}_j \cap \mathcal{M}_j = \{0\}, \tag{2}$$

$$\|k_j - P_{\mathcal{K}_0 \vee \mathcal{K}_1 \vee \dots \vee \mathcal{K}_j} k_j\| \leq 2^{-j}, \tag{3}$$

where $\mathcal{K}_j = \bigvee\{r(T)x_j : r \in R(\Omega)\}$. Assume that x_j and $\mathcal{M}_j$ have already been defined for $j < n$; note that if $n = 0$, only $\mathcal{M}_{-1}$ has been constructed and there is no f_{-1}. A repeated application of (2) yields

$$H = \mathcal{M}_{-1} = \mathcal{K}_0 \vee \mathcal{M}_0 = \mathcal{K}_0 \vee \mathcal{K}_1 \vee \mathcal{M}_1 = \cdots = \mathcal{K}_0 \vee \mathcal{K}_1 \vee \cdots \vee \mathcal{K}_{n-1} \vee \mathcal{M}_{n-1},$$

so that we can find vectors $u_n \in \mathcal{K}_0 \vee \mathcal{K}_1 \vee \cdots \vee \mathcal{K}_{n-1}$ and $v_n \in \mathcal{M}_{n-1}$ such that

$$\|k_n - u_n - v_n\| \leq 2^{-n-1}. \tag{4}$$

Choose x_n to be a $T_{|\mathcal{M}_{n-1}}$-maximal vector such that

$$\|v_n - x_n\| \leq 2^{-n-1}. \tag{5}$$

So $m_{T|\mathcal{M}_{n-1}} \equiv m_{x_n}$. We consider $T_{|\mathcal{M}_{n-1}}$ and $x_n \in \mathcal{M}_{n-1}$ and apply the splitting theorem. This shows the existence of an $R(\Omega)$-subspace $\mathcal{M}_n$ for $T_{|\mathcal{M}_{n-1}}$ satisfying (2) for $j = n$. Clearly $\mathcal{M}_n$ is $R(\Omega)$-invariant for T. Since (1) is satisfied for $j = n$ by the choice of x_n, it remains to verify (3). But since by (4) and (5)

$$\|k_n - P_{\mathcal{K}_0 \vee \dots \vee \mathcal{K}_n} k_n\| \leq \|k_n - u_n - x_n\| \leq \|k_n - u_n - v_n\| + \|v_n - x_n\| \leq 2^{-n},$$

then the existence of $\{x_j : j \geq 0\}$ and $\{\mathcal{M}_j : j \geq 0\}$ is proved by induction. An important consequence of (3) is that

$$H = \bigvee_{j<\omega} \mathcal{K}_j. \tag{6}$$

We now define a model function $\Theta = \{\theta_j : j < \omega\}$ by setting $\theta_j \equiv m_{x_j}$ for $j < \omega$. By Proposition 4.2.2, there exists a quasiaffinity X_j such that $X_j S(\theta_j) = T_{|\mathcal{K}_j} X_j$, $j < \omega$. We can then define an operator X such that $XS(\Theta) = TX$ by:

$$X(\oplus_{j<\omega} g_j) = \sum_{j<\omega} \Big(\frac{2^{-j}}{\|X_j\|}\Big) X_j g_j,$$

where

$$\oplus_{j<\omega} g_j \in \mathcal{H}(\Theta) = \bigoplus_{j<\omega} \mathcal{H}(\theta_j).$$

This formula defines a bounded operator X. Relation (6) implies that X has dense range, so that X is a quasiaffinity if we can prove that $\ker X = \{0\}$.

Assume that the vector $g = \oplus_{j<\omega} g_j \in \ker X$, $g \neq 0$ and g_n is the first nonzero component of g. We have by the definition of X that

$$X_n g_n = - \sum_{1\leq j<\omega} \|X_n\| \Big(\frac{2^{-j}}{\|X_{n+j}\|}\Big) X_{n+j} g_{n+j}. \tag{7}$$

The left-hand side of (7) belongs to $\mathcal{K}_n$, while the right-hand side of (7) belongs to $\bigvee_{1\leq j<\omega} \mathcal{K}_{n+j} \subset \mathcal{M}_n$. But by (2) $\mathcal{K}_n \cap \mathcal{M}_n = \{0\}$, so that $X_n g_n = 0$ and $g_n = 0$ since X_n is one-to-one. This contradiction implies that $g = 0$ and thus X is a quasiaffinity.

So far we proved the existence of a Jordan operator $S(\Theta)$ such that $S(\Theta) \prec T$. We can apply the same argument to T^*, and, in view of the fact that adjoints of Jordan operators are similar to Jordan operators, we deduce the existence of a Jordan operator $S(\Theta')$ such that $T \prec S(\Theta')$. The theorem now follows from the uniqueness property of Jordan operators (Theorem 4.3.9). □

COROLLARY 4.3.11. *Let T be a C_0 operator acting on the separable space H. Assume that the vectors $\{x_j : j < \omega\}$ and the rationally invariant subspaces $\{\mathcal{M}_j : j < \omega\}$ satisfy (1) and (2) in the proof of the preceding theorem with $\mathcal{M}_{-1} = H$. If $\theta_j = m_{x_j}$, $j < \omega$, then $\bigoplus_{j<\omega} S(\theta_j)$ is the Jordan model of T.*

PROOF. Fix a natural number n and set $x'_j = x_j$, and $\mathcal{M}'_j = \mathcal{M}_j$ for $0 \leq j < n$. Following the argument of the proof of Theorem 4.3.10, we can extend this finite sequences to infinite sequences $\{x'_j : j < \omega\}$ and $\{\mathcal{M}'_j : j < \omega\}$ satisfying (1) and (2) for all j and (3) for $j \geq n$ (with x_j and $\mathcal{M}_j$ replaced by x'_j and $\mathcal{M}'_j$, of course). Then the relation $H = \bigvee_{j<\omega} \mathcal{K}'_j$, $\mathcal{K}'_j = \bigvee\{r(T)x'_j : r \in r(\Omega)\}$ follows at once, and this is what needed to prove that $S(\Theta) \prec T$, where $\Theta = \{\theta'_j : j < \omega\}$ is the model function defined by $\theta'_j \equiv m_{x'_j}$, $j < \omega$. Recalling now that $x_j = x'_j$ for $j < n$, we conclude that the Jordan model $S(\Theta)$ of T satisfies the relations $\theta'_j \equiv m_{x_j}$, for $j < n$. Since n was arbitrary, the corollary follows. □

COROLLARY 4.3.12. *Let T and T' be separably acting operators of class C_0, and let $S(\Theta) = \bigoplus_{j<\omega} S(\theta_j)$ be the Jordan model of T. If $m_{T'}|\theta_j$ for every $j < \omega$, then $S(\Theta)$ is also the Jordan model of $T \oplus T'$.*

PROOF. Assume that $T \in \mathcal{L}(H)$, $T' \in \mathcal{L}(H')$ and the vectors $\{x_j : j < \omega\}$ and the $R(\Omega)$-invariant subspaces $\{\mathcal{M}_j : j < \omega\}$ satisfy the relations (1),(2) and (3) with $\mathcal{M}_{-1} = H$. If we set $\mathcal{M}'_j = \mathcal{M}_j \oplus H'$, then we have

$$m_{(T\oplus T')|\mathcal{M}'_j} \equiv m_{T|\mathcal{M}_j} \wedge m_{T'} \equiv \theta_{j+1} \wedge m_{T'} \equiv \theta_{j+1} \equiv m_{f_{j+1}}$$

for $j < \omega$, and consequently the vectors $\{x_j \oplus 0 : j < \omega\}$ and subspaces $\{\mathcal{M}'_j : j < \omega\}$ satisfy relations (1) and (2) with $\mathcal{M}_{-1} = H \oplus H'$, $\mathcal{M}_j$ replaced by $\mathcal{M}'_j$, and T replaced by $T \oplus T'$. It suffices then to apply Corollary 4.3.11 to $T \oplus T'$. □

The case of operators acting on non-separable spaces requires more details and some set-theoretical reasonings.

LEMMA 4.3.13. *Let $T \in \mathcal{L}(H)$ be an operator of class C_0. There exists a reducing subspace H_0 for T with the following property: if $\bigoplus_{j<\omega} S(\theta_j)$ is the Jordan model of $T_{|H_0}$, then $m_{T_{|H\ominus H_0}}|\theta_j$ for all $j < \omega$.*

PROOF. We apply once again the technique in the proof of Theorem 4.3.10, to construct sequences $\{x_j : j < \omega\}$ and $\{\mathcal{M}_j : j < \omega\}$ satisfying relations (1) and (2) in the proof of the same theorem, with $\mathcal{M}_{-1} = H$. We now define H_0 as the smallest reducing subspace for T containing all the vectors $\{x_j : j < \omega\}$. Let us apply Corollary 4.3.11 to $T_{|H_0}$, the vectors $\{x_j : j < \omega\}$ and the subspaces $\{\mathcal{M}'_j : j < \omega\}$, where $\mathcal{M}'_j = \mathcal{M}_j \cap H_0 = \mathcal{M}_j \ominus (H \ominus H_0)$. We deduce that the Jordan model of $T_{|H_0}$ is $\bigoplus_{j<\omega} S(\theta_j)$, where $\theta_j \equiv m_{x_j}$ for $j < \omega$. Finally we have $H \ominus H_0 \subseteq \mathcal{M}_j$ and therefore $m_{T_{|H\ominus H_0}} | m_{T_{|\mathcal{M}_{j-1}}} \equiv \theta_j$ for $j < \omega$. □

We recall that an ordinal α is a *limit ordinal* if it is not the immediate successor of another ordinal. Some examples of limit ordinals are $0, \omega, 2\omega$, and every uncountable cardinal. Every ordinal α can be uniquely written as $\alpha = \beta + k$ with β a limit ordinal and $k < \omega$.

PROPOSITION 4.3.14. *Let $T \in \mathcal{L}(H)$ be an operator of class C_0. We can associate with each limit ordinal α a reducing subspace H_α for T such that:*
(i) $H_\alpha \perp H_\beta$ if $\alpha \neq \beta$ and $H = \bigoplus_\alpha H_\alpha$;
(ii) if the Jordan model of $T_{|H_\alpha}$ is $\bigoplus_{i<\omega} S(\theta_{\alpha+i})$, then $\theta_{\alpha+i} | \theta_{\beta+j}$ whenever $\alpha > \beta$ and $i, j < \omega$.

PROOF. We use transfinite induction to associate with each limit ordinal γ reducing subspaces H_γ and $\mathcal{M}_\gamma$ for T with the following properties:

$$H_\gamma \quad \text{is separable,} \tag{1}$$

$$\Big(\bigoplus_{\beta\leq\gamma} H_\beta\Big) \oplus \mathcal{M}_\gamma = H, \tag{2}$$

$$m_{T_{|\mathcal{M}_\gamma}} | \theta_{\gamma+k} \quad \text{for} \quad k < \omega, \tag{3}$$

$$\theta_{\gamma+k} | \theta_{\beta+j} \quad \text{for} \quad \beta < \gamma \quad \text{and} \quad i, j < \omega, \tag{4}$$

where we have denoted by $\bigoplus_{j<\omega} S(\theta_{\beta+j})$ the Jordan model of $T_{|H_\beta}$.

For $\gamma = 0$ let H_0 be the space given by Lemma 4.3.13 and set $\mathcal{M}_0 = H \ominus H_0$. Now let α be a limit ordinal and assume that the spaces H_γ and $\mathcal{M}_\gamma$ have been constructed to satisfy (1),(2),(3) and (4) for all $\gamma < \alpha$. In order to construct H_α and $\mathcal{M}_\alpha$, we note that (2) implies the equality

$$\Big(\bigoplus_{\beta<\alpha} H_\beta\Big) \oplus \Big(\bigcap_{\beta<\alpha} \mathcal{M}_\beta\Big) = H.$$

We can then apply Lemma 4.3.13 once more to the operator $T_{|(\cap_{\beta<\alpha}\mathcal{M}_\beta)}$, to find a separable reducing subspace $H_\alpha \subset \cap_{\beta<\alpha}\mathcal{M}_\beta$ such that, setting $\mathcal{M}_\alpha = (\cap_{\beta<\alpha}\mathcal{M}_\beta) \ominus H_\alpha$, condition (3) is satisfied for $\gamma = \alpha$. Properties (1) and (2) are clearly satisfied for $\gamma = \alpha$, so we have only to verify (4) for $\gamma = \alpha$. If $\beta < \alpha$ we have $H_\alpha \subset \cap_{\alpha'<\alpha}\mathcal{M}_{\alpha'} \subset \mathcal{M}_\beta$ and therefore $\theta_{\alpha+i} | m_{T_{|\mathcal{M}_\beta}}$, $i < \omega$. But then we apply (3) for $\gamma = \beta$ to conclude that $\theta_{\alpha+i} | \theta_{\beta+j}$ for $i, j < \omega$. The existence of H_α and $\mathcal{M}_\alpha$ then follows by induction; (4) is in fact a consequence of the previous two conditions. Property (2) shows that $H_{\gamma+\omega} \subset \mathcal{M}_\gamma$ and therefore $\mathcal{M}_\gamma \neq 0$ if $H_{\gamma+\omega} \neq \{0\}$. Since the number of non-zero subspaces H_α cannot exceed $\dim(H)$, we conclude that $\mathcal{M}_\gamma = 0$ for some γ, and for such γ relation (2) implies that $\bigoplus_\alpha H_\alpha = H$. $\square$

COROLLARY 4.3.15. *Let $T \in \mathcal{L}(H)$ be an operator of class C_0. There exists a family of inner functions $\{\theta_\alpha : \alpha < \gamma\}$ indexed by a segment of the ordinal numbers such that:*
(i) T is quasisimilar to $\bigoplus_\alpha S(\theta_\alpha)$; and
(ii) $\theta_\alpha | \theta_\beta$ whenever $\alpha \geq \beta$.

PROOF. If H_α and $\theta_{\alpha+j}$ are given by Proposition 4.3.14, the operator T is clearly quasisimilar to $\bigoplus_\beta S(\theta_\beta)$, where the sum is extended over all ordinals β for which θ_β is not trivial; these ordinals form a segment of the ordinals by property (ii) of that proposition. The corollary now becomes obvious. $\square$

The operator $\bigoplus_\alpha S(\theta_\alpha)$ in the corollary is not necessarily a Jordan operator and hence it is not the Jordan model of T. Moreover it is not uniquely determined by T. Our task is to perform some surgery on the operator $\oplus_\alpha S(\theta_\alpha)$ given by Corollary 4.3.15, and show that this operator is quasisimilar to a Jordan operator. Let us fix an operator $T_0 = \bigoplus_\alpha S(\theta_\alpha)$, where $\{\theta_\alpha : \alpha < \gamma\}$ are such that $\theta_\beta | \theta_\alpha$ whenever $\alpha \leq \beta$.

DEFINITION 4.3.16. An ordinal number α is said to be *good* if we have $\theta_\beta \equiv \theta_\alpha$ whenever $\beta \geq \alpha$ and $\operatorname{card}(\beta) = \operatorname{card}(\alpha)$. The remaining ordinals will be called *bad.*

Of course the notion of good and bad is relative to the operator T_0. It is also quite clear that T_0 is a Jordan operator if and only if all ordinals are good. Thus we must try to eliminate the direct summands of T_0 corresponding with bad ordinals. The finite ordinals are good because there is only one ordinal with given finite cardinal, and it is not a priori obvious that other good ordinals exist.

LEMMA 4.3.17. *The set A of (equivalence classes of) inner functions defined by $A = \{\theta : \theta \equiv \theta_\alpha$ for some $\alpha\}$ is at most countable.*

PROOF. By Theorem 2.2.11 for all α we can write $\theta_\alpha \equiv B_{\mu_\alpha} S_{\nu_\alpha}$. Choose $z_0 \in \Omega$ such that $\theta_0(z_0) \neq 0$; then, since $\theta_\alpha | \theta_0$ for all α, we have that $\theta_\alpha(z_0) \neq 0$ too. Let us consider the map f defined by $\alpha \to u_\alpha(z_0)$, where u_α is the subharmonic function of Remark 2.3.7

$$u_\alpha(z) = -\sum_{\zeta \in \Omega} \mu_\alpha(\zeta) g(z, \zeta) + \int_\Gamma \frac{\partial g}{\partial n}(\zeta, z) d\nu_\alpha(\zeta).$$

If $\alpha < \beta$ then $\theta_\alpha | \theta_\beta$, and therefore by Corollary 2.3.8 we have that $u_\alpha(z) \geq u_\beta(z)$; moreover from Proposition 2.3.6 (ii) we deduce that $\mu_\alpha \leq \mu_\beta$ and $\nu_\alpha \leq \nu_\beta$. Hence $f(\alpha) \geq f(\beta)$ and, if $f(\alpha) = f(\beta)$, then $u_\alpha(z) = u_\beta(z)$ in Ω by the Maximum Modulus Principle; then $\mu_\alpha = \mu_\beta$ and $\nu_\alpha = \nu_\beta$, and thus $\theta_\alpha \equiv \theta_\beta$ by uniqueness of factorization. It follows that the set A has the same cardinality as the set $C = \{f_\alpha(z_0)\}_\alpha$. Now, we saw that the mapping $\alpha \to f_\alpha(z_0)$ is increasing and therefore C is a well ordered set. The conclusion follows from the fact that every well ordered set of real numbers with the natural order is at most countable. This last assertion is proved as follows. For every element $x \in C$ we choose a rational number $f(x)$ such that $x < f(x) < x'$ if x has an immediate successor x', and $x < f(x)$ if x is the last element of C. Thus C is equipotent with a set of rational numbers. $\square$

We can now show that there is a plentiful supply of good ordinals. We remind that θ_α are defined only for $\alpha < \gamma$, where γ can be taken to be a sufficiently large cardinal.

LEMMA 4.3.18. *For each transfinite cardinal $\aleph$ there exist good ordinals α with card(α) $= \aleph$. Moreover, if $g(\aleph) = \min\{\alpha : \text{card}(\alpha) = \aleph, \alpha \text{ is good}\}$, then the set $\{\aleph : g(\aleph) \neq \aleph\}$ is at most countable.*

PROOF. Let $\aleph$ be a transfinite cardinal and denote by $\aleph'$ the successor of $\aleph$ in the series of cardinals; of course $\aleph' > \omega$. The set of ordinals $\{\alpha : \text{card}(\alpha) = \aleph\}$ has cardinality $\aleph'$, and we have $\{\alpha : \text{card}(\alpha) = \aleph\} = \bigcup_{\theta \in A}\{\alpha : \text{card}(\alpha) = \aleph, \theta_\alpha \equiv \theta\}$, where A is the countable set of Lemma 4.3.17. The sets $\{\alpha : \text{card}(\alpha) = \aleph, \theta_\alpha \equiv \theta\}$ cannot all have cardinality $\leq \aleph$ because $\aleph_0\aleph = \aleph < \aleph'$. Thus there must exist a function $\theta \in A$ such that the set $\Sigma = \{\alpha : \text{card}(\alpha) = \aleph, \theta_\alpha \equiv \theta\}$ has cardinality $\aleph'$. Moreover that every element α of Σ is a good ordinal.

Finally, if we suppose that $\aleph_1$ and $\aleph_2$ are two distinct cardinals such that $g(\aleph_1) > \aleph_1$ and $g(\aleph_2) > \aleph_2$, then $\theta_{\aleph_1} \not\equiv \theta_{\aleph_2}$. So the mapping $\aleph \to \theta_\aleph$ is one-to-one from the set $\{\aleph : \aleph \neq g(\aleph)\}$ to the set A in Lemma 4.3.17. Thesis follows from the lemma. $\square$

REMARK 4.3.19. The preceding proof shows actually that the set

$$\{\alpha : \text{card}(\alpha) = \aleph, \ \alpha \text{ is good}\} = \{\alpha : g(\aleph) \leq \alpha < \aleph'\}$$

has cardinality $\aleph'$, while the set of bad ordinals $\{\alpha : \text{card}(\alpha) = \aleph, \aleph \leq \alpha < g(\aleph)\}$ has cardinality $\leq \aleph$. We can in fact construct a bijection of all ordinals onto the good ordinals. Indeed, we recall that if two ordinals α and β satisfy the relation $\alpha < \beta$, then β can be written uniquely as $\beta = \alpha + \gamma$. The number γ defined in this way will be denoted $\beta - \alpha$. Then the bijection mentioned above can be obtained by associating with each ordinal α the good ordinal $g(\text{ card}(\alpha)) + (\alpha - \text{ card}(\alpha))$. As a consequence, the operator $\bigoplus_{\alpha \text{ is good}} S(\theta_\alpha)$ is unitarily equivalent to the Jordan operator $S(\Theta')$ where $\Theta' = \{\theta'_\alpha\}$ is given by $\theta'_\alpha = \theta_{g(\text{ card}(\alpha))}$.

We prove now the last result that we need to eliminate all summands corresponding with bad ordinals.

LEMMA 4.3.20. *There exists a function f defined on the set of bad ordinal numbers such that:*
(i) $f(\alpha) <$ card(α) is a good limit ordinal; and
(ii) $f(\alpha) \neq f(\beta)$ if $\alpha \neq \beta$ and card(α) $=$ card(β) $> \aleph_0$.
If f is any such function, then $f(\alpha) = 0$ whenever $\aleph_0 \leq \alpha < g(\aleph_0)$ and the set $\{\alpha : f(\alpha) = \beta\}$ is at most countable for every limit ordinal β.

PROOF. If $\aleph_0 \leq \alpha < g(\aleph_0)$, then 0 is the only good limit ordinal that is $\leq \alpha$ and therefore we define $f(\alpha) = 0$. If $\aleph$ is an uncountable ordinal, using the above remarks, we have that $\{\alpha : \alpha < \aleph,\ \alpha \text{ is good}\}$ has cardinality $\aleph$ and thus $\Sigma_\aleph = \{\alpha : \alpha < \aleph, \alpha \text{ is a good limit ordinal}\}$ also has cardinality $\aleph$. Since the set of bad ordinals $\Omega_\aleph = \{\alpha : \aleph \leq \alpha < g(\aleph)\}$ has cardinality $\leq \aleph$, we can define f such that $f(\Omega_\aleph) \subset \Sigma_\aleph$ and $f_{|\Omega_\aleph}$ is one-to-one. To finish the proof we need only to check that $\{\alpha : f(\alpha) = \beta\}$ is at most countable. By (ii), for each uncountable $\aleph$ such that $\aleph \neq g(\aleph)$ there exists at most one α such that $\operatorname{card}(\alpha) = \aleph$ and $f(\alpha) = \beta$. Hence the lemma follows from the fact that the sets $\{\aleph : \aleph > \aleph_0, g(\aleph) \neq \aleph\}$ and $\{\alpha : \aleph_0 \leq \alpha < g(\aleph_0)\}$ are at most countable. $\square$

THEOREM 4.3.21. **Classification Theorem** *Every operator T of class C_0 is quasisimilar to a Jordan operator $S(\Theta)$. Moreover $S(\Theta)$ is uniquely determined by either of the relations $S(\Theta) \prec T$ and $T \prec S(\Theta)$.*

PROOF. The uniqueness follows as in the proof of Theorem 4.3.10. For the existence part we may assume that $T = \bigoplus_\alpha S(\theta_\alpha)$, where the inner functions $\{\theta_\alpha\}$ can be chosen as in Corollary 4.3.15. As we saw in Remark 4.3.19, the operator $T' = \bigoplus_{\alpha \text{ is good}} S(\theta_\alpha)$ is unitarily equivalent to a Jordan operator. Therefore it will suffice to show that T and T' are quasisimilar. Let f be the function on bad ordinals given by the last lemma and note that T is unitarily equivalent to

$$T'' = T''' \oplus \Big\{ \bigoplus_{\alpha \text{ is a good limit ordinal}} \Big[\big(\bigoplus_{j<\omega} S(\theta_{\alpha+j})\big) \oplus \big(\bigoplus_{f(\beta)=\alpha} S(\theta_\beta)\big)\Big]\Big\},$$

where T''' collects all good ordinals that do not have the form $\alpha + j$ with a good limit ordinal α and $j < \omega$. Of course $\alpha + j$ is good whenever α is good and $j < \omega$. Now, we know that the relation $f(\beta) = \alpha$ implies that $\operatorname{card}(\beta) \geq \aleph_0$ and $\alpha \leq \operatorname{card}(\beta)$. Consequently $\alpha + j \leq \beta$ for all $j < \omega$, and this implies that $\theta_\beta | \theta_{\alpha+j}$ if $f(\beta) = \alpha$ and $j < \omega$. Therefore the minimal function of the separably acting operator $\bigoplus_{f(\beta)=\alpha} S(\theta_\beta)$ divides $\theta_{\alpha+j}$ for all $j < \omega$, and Corollary 4.3.12 implies that

$$\Big(\bigoplus_{j<\omega} S(\theta_{\alpha+j})\Big) \oplus \Big(\bigoplus_{f(\beta)=\alpha} S(\theta_\beta)\Big)$$

is quasisimilar to $\bigoplus_{j<\omega} S(\theta_{\alpha+j})$. We conclude that T'' is quasisimilar to

$$T''' \oplus \Big\{ \bigoplus_{\alpha \text{ is a good limit ordinal}} \big(\bigoplus_{j<\omega} S(\theta_{\alpha+j})\big)\Big\},$$

while the last operator is unitarily equivalent to T'. The theorem follows. $\square$

COROLLARY 4.3.22. *For every operator T of class C_0 we have $\mu_T = \mu_{T^*}$.*

PROOF. Let $S(\Theta)$ be the Jordan model of T. By Lemma 4.1.9 we have $\mu_T = \mu_{S(\Theta)}$ and $\mu_{T^*} = \mu_{S(\Theta)^*}$, and the equality $\mu_{S(\Theta)} = \mu_{S(\Theta)^*}$ follows by Theorem 4.3.7. $\square$

COROLLARY 4.3.23. *If T is of class C_0 and $\mathcal{M}$ is an $R(\Omega)$-invariant subspace for T, then $\mu_{T_{|\mathcal{M}}} \leq \mu_T$.*

PROOF. We have $(T_{|\mathcal{M}})^* P_{\mathcal{M}} = P_{\mathcal{M}} T^*$ and by Proposition 4.1.11 $\mu_{(T|\mathcal{M})^*} \leq \mu_{T^*}$. The result follows then from the preceding corollary. □

There are many consequences of the classification theorem. We note a description of the $R(\Omega)$-invariant subspaces of an operator of class C_0, whose proof is similar to the proof of Proposition 3.5.33 in [7].

PROPOSITION 4.3.24. *Let T be an operator of class C_0 and $\mathcal{M}$ be an $R(\Omega)$-invariant subspace for T. There exist operators X and Y in $\{T\}'$ such that* $\mathcal{M} = (\, ranX)^- = \ker Y$.

If T is a linear operator on a finite-dimensional space H, then the classical theorem of Jordan shows that T is similar to a direct sum of Jordan cells. In particular, H can be decomposed into a direct (not orthogonal) sum of invariant subspaces for T such that the restriction of T to each of these subspaces is similar to some Jordan cell. A weaker notion of decomposition gives us analogues of the finite-dimensional result. The space on which an operator of class C_0 acts can be decomposed into rationally cyclic invariant subspaces. The kind of decomposition that can be obtained is somewhat weaker than a direct sum decomposition. This parallels the fact that quasisimilarity is a weaker relation than similarity.

DEFINITION 4.3.25. Let $\{H_j : j \in J\}$ be a family of closed subspaces of H satisfying the relation $H = \vee_{j \in J} H_j$. We say that H is the *almost direct sum* of the family $\{H_j : j \in J\}$ if for every family $\{K_\alpha : \alpha \in A\}$ of subsets of J such that $\cap_{\alpha \in A} K_\alpha = \emptyset$ we have $\cap_{\alpha \in A}(\vee_{j \in K_\alpha} H_j) = \{0\}$.

THEOREM 4.3.26. *Let $T \in \mathcal{L}(H)$ be an operator of class C_0, and let $S(\Theta)$, $\Theta = \{\theta_\alpha\}$, be the Jordan model of T. We can associate with each ordinal α an $R(\Omega)$-invariant subspace H_α for T with the following properties:*
(i) H is the almost direct sum of $\{H_\alpha\}$;
(ii) $T_{|H_\alpha}$ is quasisimilar to $S(\theta_\alpha)$ for each α; and
(iii) $H_{\alpha+i} \perp H_{\beta+j}$ if α and β are different limit ordinals and $i, j < \omega$.

The proof of this theorem, that we omit, is a refinement of the proof of Theorem 4.3.10, using a more complicate splitting principle similar to Theorem 4.3.6 in [7]. The first step of the proof consists in reducing to the separable case by means of a result similar to Theorem 3.6.4 in [7].

Bibliography

[1] M. B. Abrahamse, *Toeplitz Operators in multiply connected regions,* Amer. J. Math. **96** (1974), 261-297.

[2] J. Agler, *Rational dilation on an annulus,* Annals of Math. **121** (1985), 537-563.

[3] A. Aleman and S. Richter, *Nearly and Simply Invariant Subspaces of H^2 of some Multiply Connected Regions,* Preprint.

[4] M. B. Abrahamse and R. G. Douglas, *Operators on multiply connected domains,* Proc. of the Royal Acad. **74 A** (1974), 135-141.

[5] M. B. Abrahamse and R. G. Douglas, *A Class of Subnormal Operators related to multiply connected domains,* Advances in Math. **19** (1976), 106-148.

[6] J. A. Ball, *Operators of Class C_{00} over Multiply-Connected Domains,* Mich. Math. J. **25** (1978), 183-196.

[7] H. Bercovici, *Operator Theory and Arithmetic in H^∞,* Amer. Math. Soc., Providence, Rhode Island (1988).

[8] H. Bercovici and W. S. Li, *Normal Boundary Dilations and rationally invariant subspaces,* Integr. Equat. Oper. Th. **15** (1992), 709-721.

[9] H. Bercovici and A. Zucchi, *Generalized Interpolation in a Multiply Connected Region,* to appear on Proc. Amer. Math. Soc.

[10] A. Beurling, *On two problems concerning linear transformations in Hilbert space,* Acta Math. **81** (1949), 239-255.

[11] B. Chevreau, C. M. Pearcy, A. L. Shields, *Finitely connected domains G, representations of $H^\infty(G)$, and invariant subspaces,* J. Operator Theory **6** (1981), 375-405.

[12] R. G. Douglas and V. Paulsen, *Completely bounded maps and Hypo-Dirichlet Algebras,* Acta Sa. Math. (Szeged) **50** (1986), 143-157.

[13] S. Fisher, *Function Theory on Planar Domains, a second course in Complex Analysis,* Wiley, New York (1983).

[14] T. Gamelin and G. Garnett, *Pointwise bounded approximation and Dirichlet Algebras,* J. Functional Analysis **8** (1971), 360-404.

[15] G. M. Goluzin, *Geometric Theory of Function of a Complex Variable,* (Moscow, 1952). English transl.: Amer. Math. Soc., Providence, Rhode Island (1969).

[16] M. Hasumi, *Invariant subspace theorems for finite Riemann surfaces,* Canad. J. Math. **18** (1966), 240-255.

[17] R. V. Kadison and J. R. Ringrose, *Fundamentals of the theory of operator algebras, Vol. I,* Academic Press,(1983).

[18] D. Khavinson, *Factorization Theorems for different Classes of Analytic Functions in Multiply Connected Domains,* Pacific J. Math. **108** (1983), 295-318.

[19] D. Khavinson, *On Removal of periods of conjugate functions in multiply connected domains,* Mich. Math. J. **31** (1984), 371-379.

[20] H. L. Royden, *Invariant subspaces of H^p for multiply connected regions,* Pacific J. Math. **134** (1988), 151-172.
[21] W. Rudin, *Analytic functions of class H_p*, Trans. Amer. Math. Soc. **78** (1955), 46-66.
[22] D. Sarason, *The H^p spaces of an annulus,* Memoirs of the Amer. Math. Soc. **56**, Providence, Rhode Island (1965).
[23] D. Sarason, *Generalized Interpolation in H^∞*, Trans. Amer. Math. Soc. **127** (1967), 179-203.
[24] B. Sz.-Nagy, *Sur les contractions de l'espace de Hilbert,* Acta Sci Math. (Szeged) **15** (1953), 87-92.
[25] B. Sz.-Nagy, *Sur les contractions de l'espace de Hilbert II,* Acta Sci Math. (Szeged) **18** (1957), 1-15.
[26] B. Sz.-Nagy and C. Foias, *Sur les contractions de l'espace de Hilbert VII. Triangulations canoniques, functions minimum,* Acta Sci. Math. (Szeged) **25** (1964), 12-37.
[27] B. Sz.-Nagy and C. Foias, *Harmonic Analysis of operators on Hilbert space,* North-Holland, Amsterdam (1970).
[28] J. von Neumann, *Allgemrine Eigenwerttheorie Hermitescher Funktional-operatoren,* Math. Annln. **102** (1929), 49-131.
[29] J. von Neumann, *Eine Spektral theorie für allgemeine Operatoren eines unitären Raumes,* Math. Nachr. **4** (1951), 258-281.
[30] M. Voichick, *Ideals and invariant subspaces of analytic functions,* Trans. Amer. Math. Soc. **111** (1964), 493-512.
[31] M. Voichick, *Invariant subspaces on Riemann surfaces,* Canad. J. Math. **18** (1966), 399-403.
[32] H. Wold, *A study in the analysis of stationary time series,* Stockholm, 1938; 2nd ed., 1954.
[33] A. Zucchi, Ph.D. Dissertation, Indiana University, Bloomington (1994).

Editorial Information

To be published in the *Memoirs*, a paper must be correct, new, nontrivial, and significant. Further, it must be well written and of interest to a substantial number of mathematicians. Piecemeal results, such as an inconclusive step toward an unproved major theorem or a minor variation on a known result, are in general not acceptable for publication. *Transactions* Editors shall solicit and encourage publication of worthy papers. Papers appearing in *Memoirs* are generally longer than those appearing in *Transactions* with which it shares an editorial committee.

As of January 31, 1997, the backlog for this journal was approximately 7 volumes. This estimate is the result of dividing the number of manuscripts for this journal in the Providence office that have not yet gone to the printer on the above date by the average number of monographs per volume over the previous twelve months, reduced by the number of issues published in four months (the time necessary for preparing an issue for the printer). (There are 6 volumes per year, each containing at least 4 numbers.)

A Copyright Transfer Agreement is required before a paper will be published in this journal. By submitting a paper to this journal, authors certify that the manuscript has not been submitted to nor is it under consideration for publication by another journal, conference proceedings, or similar publication.

Information for Authors and Editors

Memoirs are printed by photo-offset from camera copy fully prepared by the author. This means that the finished book will look exactly like the copy submitted.

The paper must contain a *descriptive title* and an *abstract* that summarizes the article in language suitable for workers in the general field (algebra, analysis, etc.). The *descriptive title* should be short, but informative; useless or vague phrases such as "some remarks about" or "concerning" should be avoided. The *abstract* should be at least one complete sentence, and at most 300 words. Included with the footnotes to the paper, there should be the 1991 *Mathematics Subject Classification* representing the primary and secondary subjects of the article. This may be followed by a list of *key words and phrases* describing the subject matter of the article and taken from it. A list of the numbers may be found in the annual index of *Mathematical Reviews*, published with the December issue starting in 1990, as well as from the electronic service e-MATH [**telnet e-MATH.ams.org** (or **telnet 130.44.1.100**). Login and password are **e-math**]. For journal abbreviations used in bibliographies, see the list of serials in the latest *Mathematical Reviews* annual index. When the manuscript is submitted, authors should supply the editor with electronic addresses if available. These will be printed after the postal address at the end of each article.

Electronically prepared papers. The AMS encourages submission of electronically prepared papers in AMS-TeX or AMS-LaTeX. The Society has prepared author packages for each AMS publication. Author packages include instructions for preparing electronic papers, the *AMS Author Handbook*, samples, and a style file that generates the particular design specifications of that publication series for both AMS-TeX and AMS-LaTeX.

Authors with FTP access may retrieve an author package from the Society's Internet node `e-MATH.ams.org` (130.44.1.100). For those without FTP access, the author package can be obtained free of charge by sending e-mail to `pub@math.ams.org` (Internet) or from the Publication Division, American Mathematical Society, P.O. Box 6248, Providence, RI 02940-6248. When requesting an author package, please specify $\mathcal{A}_{\mathcal{M}}\mathcal{S}$-TEX or $\mathcal{A}_{\mathcal{M}}\mathcal{S}$-LATEX, Macintosh or IBM (3.5) format, and the publication in which your paper will appear. Please be sure to include your complete mailing address.

Submission of electronic files. At the time of submission, the source file(s) should be sent to the Providence office (this includes any TEX source file, any graphics files, and the DVI or PostScript file).

Before sending the source file, be sure you have proofread your paper carefully. The files you send must be the EXACT files used to generate the proof copy that was accepted for publication. For all publications, authors are required to send a printed copy of their paper, which exactly matches the copy approved for publication, along with any graphics that will appear in the paper.

TEX files may be submitted by email, FTP, or on diskette. The DVI file(s) and PostScript files should be submitted only by FTP or on diskette unless they are encoded properly to submit through e-mail. (DVI files are binary and PostScript files tend to be very large.)

Files sent by electronic mail should be addressed to the Internet address `pub-submit@math.ams.org`. The subject line of the message should include the publication code to identify it as a Memoir. TEX source files, DVI files, and PostScript files can be transferred over the Internet by FTP to the Internet node `e-math.ams.org` (130.44.1.100).

Electronic graphics. Figures may be submitted to the AMS in an electronic format. The AMS recommends that graphics created electronically be saved in Encapsulated PostScript (EPS) format. This includes graphics originated via a graphics application as well as scanned photographs or other computer-generated images.

If the graphics package used does not support EPS output, the graphics file should be saved in one of the standard graphics formats—such as TIFF, PICT, GIF, etc.—rather than in an application-dependent format. Graphics files submitted in an application-dependent format are not likely to be used. No matter what method was used to produce the graphic, it is necessary to provide a paper copy to the AMS.

Authors using graphics packages for the creation of electronic art should also avoid the use of any lines thinner than 0.5 points in width. Many graphics packages allow the user to specify a "hairline" for a very thin line. Hairlines often look acceptable when proofed on a typical laser printer. However, when produced on a high-resolution laser imagesetter, hairlines become nearly invisible and will be lost entirely in the final printing process.

Screens should be set to values between 15% and 85%. Screens which fall outside of this range are too light or too dark to print correctly.

Any inquiries concerning a paper that has been accepted for publication should be sent directly to the Editorial Department, American Mathematical Society, P. O. Box 6248, Providence, RI 02940-6248.

Editors

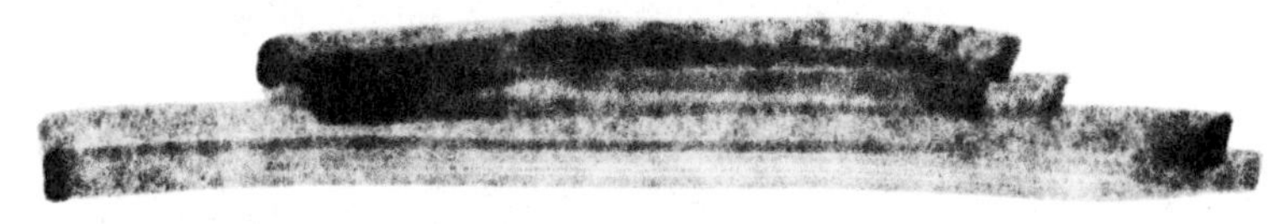

Selected Titles in This Series

(Continued from the front of this publication)

577 **Freddy Dumortier and Robert Roussarie,** Canard cycles and center manifolds, 1996

576 **Grahame Bennett,** Factorizing the classical inequalities, 1996

575 **Dieter Heppel, Idun Reiten, and Sverre O. Smalø,** Tilting in Abelian categories and quasitilted algebras, 1996

574 **Michael Field,** Symmetry breaking for compact Lie groups, 1996

573 **Wayne Aitken,** An arithmetic Riemann-Roch theorem for singular arithmetic surfaces, 1996

572 **Ole H. Hald and Joyce R. McLaughlin,** Inverse nodal problems: Finding the potential from nodal lines, 1996

571 **Henry L. Kurland,** Intersection pairings on Conley indices, 1996

570 **Bernold Fiedler and Jürgen Scheurle,** Discretization of homoclinic orbits, rapid forcing and "invisible" chaos, 1996

569 **Eldar Straume,** Compact connected Lie transformation groups on spheres with low cohomogeneity, I, 1996

568 **Raúl E. Curto and Lawrence A. Fialkow,** Solution of the truncated complex moment problem for flat data, 1996

567 **Ran Levi,** On finite groups and homotopy theory, 1995

566 **Neil Robertson, Paul Seymour, and Robin Thomas,** Excluding infinite clique minors, 1995

565 **Huaxin Lin and N. Christopher Phillips,** Classification of direct limits of even Cuntz-circle algebras, 1995

564 **Wensheng Liu and Héctor J. Sussmann,** Shortest paths for sub-Riemannian metrics on rank-two distributions, 1995

563 **Fritz Gesztesy and Roman Svirsky,** (m)KdV solitons on the background of quasi-periodic finite-gap solutions, 1995

562 **John Lindsay Orr,** Triangular algebras and ideals of nest algebras, 1995

561 **Jane Gilman,** Two-generator discrete subgroups of $PSL(2, R)$, 1995

560 **F. Tomi and A. J. Tromba,** The index theorem for minimal surfaces of higher genus, 1995

559 **Paul S. Muhly and Baruch Solel,** Hilbert modules over operator algebras, 1995

558 **R. Gordon, A. J. Power, and Ross Street,** Coherence for tricategories, 1995

557 **Kenji Matsuki,** Weyl groups and birational transformations among minimal models, 1995

556 **G. Nebe and W. Plesken,** Finite rational matrix groups, 1995

555 **Tomás Feder,** Stable networks and product graphs, 1995

554 **Mauro C. Beltrametti, Michael Schneider, and Andrew J. Sommese,** Some special properties of the adjunction theory for 3-folds in $\mathbb{P}^5$, 1995

553 **Carlos Andradas and Jesús M. Ruiz,** Algebraic and analytic geometry of fans, 1995

552 **C. Krattenthaler,** The major counting of nonintersecting lattice paths and generating functions for tableaux, 1995

551 **Christian Ballot,** Density of prime divisors of linear recurrences, 1995

550 **Huaxin Lin,** C^*-algebra extensions of $C(X)$, 1995

549 **Edwin Perkins,** On the martingale problem for interactive measure-valued branching diffusions, 1995

548 **I-Chiau Huang,** Pseudofunctors on modules with zero dimensional support, 1995

547 **Hongbing Su,** On the classification of C*-algebras of real rank zero: Inductive limits of matrix algebras over non-Hausdorff graphs, 1995

546 **Masakazu Nasu,** Textile systems for endomorphisms and automorphisms of the shift, 1995

545 **John L. Lewis and Margaret A. M. Murray,** The method of layer potentials for the heat equation on time-varying domains, 1995

544 **Hans-Otto Walther,** The 2-dimensional attractor of $x'(t) = -\mu x(t) + f(x(t-1))$, 1995

the AMS catalog for earlier titles)